WIRELESS SENSOR AND ACTOR NETWORKS

IFIP – The International Federation for Information Processing

IFIP was founded in 1960 under the auspices of UNESCO, following the First World Computer Congress held in Paris the previous year. An umbrella organization for societies working in information processing, IFIP's aim is two-fold: to support information processing within its member countries and to encourage technology transfer to developing nations. As its mission statement clearly states,

> *IFIP's mission is to be the leading, truly international, apolitical organization which encourages and assists in the development, exploitation and application of information technology for the benefit of all people.*

IFIP is a non-profitmaking organization, run almost solely by 2500 volunteers. It operates through a number of technical committees, which organize events and publications. IFIP's events range from an international congress to local seminars, but the most important are:

• The IFIP World Computer Congress, held every second year;
• Open conferences;
• Working conferences.

The flagship event is the IFIP World Computer Congress, at which both invited and contributed papers are presented. Contributed papers are rigorously refereed and the rejection rate is high.

As with the Congress, participation in the open conferences is open to all and papers may be invited or submitted. Again, submitted papers are stringently refereed.

The working conferences are structured differently. They are usually run by a working group and attendance is small and by invitation only. Their purpose is to create an atmosphere conducive to innovation and development. Refereeing is less rigorous and papers are subjected to extensive group discussion.

Publications arising from IFIP events vary. The papers presented at the IFIP World Computer Congress and at open conferences are published as conference proceedings, while the results of the working conferences are often published as collections of selected and edited papers.

Any national society whose primary activity is in information may apply to become a full member of IFIP, although full membership is restricted to one society per country. Full members are entitled to vote at the annual General Assembly, National societies preferring a less committed involvement may apply for associate or corresponding membership. Associate members enjoy the same benefits as full members, but without voting rights. Corresponding members are not represented in IFIP bodies. Affiliated membership is open to non-national societies, and individual and honorary membership schemes are also offered.

WIRELESS SENSOR AND ACTOR NETWORKS

IFIP WG 6.8 First International Conference on Wireless Sensor and Actor Networks, WSAN'07, Albacete, Spain, September 24-26, 2007

Edited by

Luis Orozco-Barbosa
Universidad de Castilla-La Mancha
Spain

Teresa Olivares
Universidad de Castilla-La Mancha
Spain

Rafael Casado
Universidad de Castilla-La Mancha
Spain

Aurelio Bermúdez
Universidad de Castilla-La Mancha
Spain

 Springer

Wireless Sensor and Actor Networks

Edited by L. Orozco-Barbosa, T. Olivares, R. Casado, and A. Bermúdez

   p.   cm. (IFIP International Federation for Information Processing, a Springer Series in Computer Science)

   ISSN: 1571-5736 / 1861-2288 (Internet)

ISBN 978-1-4419-4515-0   eISBN: 13: 978-0-387-74899-3
Printed on acid-free paper

9 8 7 6 5 4 3 2 1
springer.com

Preface

The IFIP Working Group 6.8 Mobile and Wireless Communications has a long tradition on addressing and grouping researchers and practitioners working on various mobile and wireless communications technologies and services. Due to the promising and exciting applications enabled by the development of Wireless Sensor and Actor Networks (WSAN), the IFIP WG 6.8 had decided to launch a new series of conferences on this exciting new technology. The 1st WSAN was held in Albacete, Spain on September 24-26, 2006. After a thoroughly evaluation process by the program committee members assisted by external reviewers, a total of 20 papers from 9 different countries were selected to be included in the program.

The papers selected to be included in the volume illustrate the state-of-the-art and current trends in the area of wireless sensor and actor networks. The program was organized into eight topics:

1. Actors
2. Applications
3. Security
4. Energy
5. Quality of Service
6. Localization
7. Middleware
8. Protocols

We are grateful to Dan Steignart from the University of Berkeley for having accepted to deliver the opening tutorial, and Pedro Marrón from the University of Bonn, Luis Redondo from MTP and Walter Stockwell from CrossBow for having accepted to participate in the panel session. We would like to thank all the members of the Technical

Program Committee and the additional referees. Without the support, the conference organization would not have been possible. Last but not least, we are grateful to all the authors and participants who trusted us to organize this event and to Springer's IFIP Editorial for supporting us. We expect WSAN 2007 to have been a fruitful and stimulating forum for exchanging ideas and experiences in the area of wireless sensor and actor networks.

September 2007

Luis Orozco-Barbosa
Teresa Olivares
Rafael Casado
Aurelio Bermudez

Acknowledgements

General Co-chairs

Luis OROZCO-BARBOSA   Universidad de Castilla-La Mancha, Spain
Teresa OLIVARES   Universidad de Castilla-La Mancha, Spain

Program Co-chairs

Al Agha KHALDOUN   University of Paris-Sud, France
Otto DUARTE   Universidad Federal de Río de Janeiro, Brasil

Steering Committee

Augusto CASACA   INESC, Portugal
Ramón PUIGJANER   Universidad de las Islas Baleares, Spain
Al Agha KHALDOUN   University of Paris-Sud , France
Ivan STOJMENOVIC   University of Ottawa, Canada
Luis OROZCO-BARBOSA   Universidad de Castilla La Mancha, Spain
Guy PUJOLLE   LIP6, France
Otto DUARTE   Universidad Federal de Río de Janeiro, Brasil
Teresa OLIVARES   Universidad de Castilla-La Mancha, Spain
Pedro MARRÓN   University of Sttutgart, Germany
Pedro CUENCA   Universidad de Castilla-La Mancha, Spain

Publications Chair
Rafael CASADO   Universidad de Castilla-La Mancha, Spain

Publicity Chair
Aurelio BERMÚDEZ   Universidad de Castilla-La Mancha, Spain

Technical Program Committee

Tarek ABDELZAHER	Univ. of Illinois at Urbana Champaign, USA
Muneeb ALI	TU Delft, The Netherlands
Guillermo BARRENETXEA	École Pol. Féd. de Lausanne, Switzerland
Torsten BRAUN	ETH, Switzerland
M. Ufuk CAGLAYAN	Bogazici University, Turkey
Augusto CASACA	INESC, Portugal
Marco CONTI	National Research Council, Italy
Otto DUARTE	Universidad Federal de Río de Janeiro, Brasil
Jean-Pierre EBERT	IHP microelectronics, Germany
Luigi FRATTA	Politecnico di Milano, Italy
Erol GELENBE	Imperial College, United Kingdom
Mario GERLA	UCLA, USA
Lewis GIROD	MIT, USA
Takahiro HARA	Osaka University, Japan
Qingfeng HUANG	ECCA, Palo Alto Research Center, USA
Ahmed KAMAL	Iowa State University, USA
Farouk KAMOUN	ENSI, Tunisia
Aman KANSAL	Microsoft Research, USA
Al Agha KHALDOUN	University of Paris-Sud , France
Dongkyun KIM	Kyunpook National University, Korea
Young-Bae KO	Ajou University, Korea
Miguel LOPEZ	UAM-I, Mexico
Pedro MARRÓN	University of Stttutgart, Germany
Ali MIRI	University of Ottawa, Canada
Manuel José PEREZ	Universidad Miguel Hernández, Spain
Viktor K. PRASANNA	University of Southern California, USA
Ramón PUIGJANER	Universidad de las Islas Baleares, Spain
Guy PUJOLLE	LIP6, France
Hartmut RITTER	ScatterWeb GmbH, Berlin, Germany
Pedro M. RUIZ	Universidad de Murcia, Spain
Jaime SANCHEZ	CICESE, Mexico
Ottio SPANIOL	Univ. of Technology of Aechen, Germany
Avinash SRINIVASAN	Florida Atlantic University, USA
Ivan STOJMENOVIC	University of Ottawa, Canada
Bulent TAVLI	TOBB Economy and Tech. Univ., Turkey
Kamin WHITEHOUSE	University of Virginia, USA
Andreas WILLIG	Technical University of Berlin, Germany
Eric M. YEATMAN	Imperial College London, United Kingdom

Organizing Committee

José C. CASTILLO	Universidad de Castilla-La Mancha, Spain
Francisco M. DELICADO	Universidad de Castilla-La Mancha, Spain
Jesús DELICADO	Universidad de Castilla-La Mancha, Spain
Raúl GALINDO	Universidad de Castilla-La Mancha, Spain
Eva M. GARCÍA	Universidad de Castilla-La Mancha, Spain
José L. MARTÍNEZ	Universidad de Castilla-La Mancha, Spain
Antonio M. ORTIZ	Universidad de Castilla-La Mancha, Spain
Paz PEDRÓN	Universidad de Castilla-La Mancha, Spain
M. Fátima REQUENA	Universidad de Castilla-La Mancha, Spain
Antonio ROBLES	Universidad de Castilla-La Mancha, Spain
Fernando ROYO	Universidad de Castilla-La Mancha, Spain
José VILLALÓN	Universidad de Castilla-La Mancha, Spain

Sponsoring Institutions

DSI	Departamento de Sistemas Informáticos, UCLM
EPSA	Escuela Politécnica Superior de Albacete
I3A	Instituto de Investigación en Informática
PCyTA	Parque Científico y Tecnológico de Albacete
UCLM	Universidad de Castilla-La Mancha
JCCM	Junta de Comunidades de Castilla-La Mancha
MEC	Ministerio de Educación y Ciencia

Table of contents

Actors

Applications

Security

Energy & QoS

Localization & Middleware

Protocols

LOCALIZED MOVEMENT CONTROL FOR FAULT TOLERANCE OF MOBILE ROBOT NETWORKS

Shantanu Das[1], Hai Liu[1], Ajith Kamath[1], Amiya Nayak[1], and Ivan Stojmenović[1,2]

[1] School of Information Technology and Engineering, University of Ottawa, Canada
[2] Electronic, Electrical & Computer Engineering, The University of Birmingham, United Kingdom
{shantdas,akamath,hailiu,anayak,ivan}@site.uottawa.ca

Abstract. In this paper, we present a novel localized movement control algorithm to form a fault-tolerant bi-connected robotic network topology from a connected network, such that total distance of movement of robots is minimized. The proposed distributed algorithm uses p-hop neighbor information to identify critical head robots that can direct two neighbors to move toward each other and bi-connect their neighborhood. Simulation results show that the total distance of movement of robots decreases significantly with our localized algorithm when compared to the globalized one, and our localized algorithm achieved 100% success on considered non-bi-connected networks. To the best of our knowledge, it is the first work on localized movement control for fault tolerance of mobile robot networks.

Keywords : mobile sensors, actor networks, fault tolerance, robot movement, localized control.

1 Introduction

A significant trend in the development of autonomous mobile robot networks is seen due to the advancement of sophisticated robots. Mobile robotics is expected to be a solution ranging from vast number of industrial applications to service robotics. In such applications coordination between individual robots is essentially accomplished through a wireless ad hoc network. For example, coordination of robotic relay stations was studied in [5] to maintain communication between an explorer and a base station. Application of mobile robotics is vast. Potential applications include military missions, unmanned space exploration, and data collection in sensor fields. But for such applications, coordination of a robot team in pursuit of common task is essential. Existing algorithms for mobile robots coordination are suitable for robots with no or very low failure rates. However, when robots are susceptible to failures, as in many applications, it is critical for robotic networks to incorporate the ability to sustain faults

Please use the following format when citing this chapter:

Dasi, S., Liu, H., Kamath, A., Nayak, A., Stojmenovic, I., 2007, in IFIP International Federation for Information Processing, Volume 248, Wireless Sensor and Actor Networks, eds. L. Orozco-Barbosa, Olivares, T., Casado, R., Bermudez, A., (Boston: Springer), pp. 1-12.

and operate normally. Communication faults in robot networks can be caused by hardware damage, energy depletion, harsh environment conditions and malicious attacks. A fault in a robot can cause stopping transmission tasks to others as well as relaying data to sink. Data sent by a robot will be lost if the receiving robot fails. So, a communication link failure on a route requires data to be rerouted. That is, in order to handle general communication faults, there should be at least two node-disjoint paths between each pair of robots in the network. A network is defined to be **bi-connected** if there exist two node-disjointed paths between any pair of nodes in the network, i.e., the removal of any node from the network leaves the network still connected. Therefore, bi-connectivity is the basic requirement for design of fault-tolerant networks [9].

In this paper, we focus on mobile robot networks and study movement control of robots to establish a fault-tolerant bi-connected network. The robot network is assumed to be connected, but not necessarily bi-connected. Achieving connectivity in a disconnected network is difficult due to the lack of communication between the disconnected parts. However, if the network is already connected, we can make it bi-connected (and thus fault-tolerant) by movement of selected robots. Recent work in [2] has shown that fault tolerance can be achieved through globalized robot movement control algorithm. It is a *centralized* algorithm that assumes one of robots or a base station has global information of the network. We focus on the *localized* version of movement control algorithm for building a fault-tolerant robot network. To the best our knowledge, this is the first work on localized movement control for fault tolerance of mobile robot networks.

The rest of the paper is organized as follows. Related work is introduced in Section 2. We propose a localized movement control algorithm to construct bi-connected mobile robot networks in Section 3. Results obtained from extensive simulations are provided in Section 4 to show the effectiveness of our algorithm. Finally we conclude our work in Section 5.

2 Related Work

Many topology control algorithms have been proposed to achieve network reliability in static networks. These algorithms cope with preserving fault tolerance by selecting certain links to neighbors in an already well connected network. The problem of adjusting the transmit power of nodes to create a desired topology in multiple wireless networks was studied in [10]. For static networks, two centralized algorithms were proposed to construct connected and bi-connected networks while minimizing the maximal transmission power of nodes. Two distributed heuristics were further proposed for mobile networks. The basic idea is to adaptively adjust node transmit power according to topological changes and attempt to maintain a connected topology with the minimum power. A more general case for k-vertex connectivity of wireless networks was studied in [7]. Both a centralized algorithm and a localized algorithm were proposed. Both above works assumed that nodes have uniform transmission range. That is, they focused on homogenous networks. Topology control in heterogeneous wireless networks was discussed in [7]. Two localized algorithms were proposed. It was proved that the topologies generated by the proposed algorithms preserve bi-connectivity of networks. An extension of cone-based topology control algorithm was proposed in [1]. Each node decides its own power based on local information about relative angle of its neighbors. It showed that a fault-tolerant network topology is achievable and transmission power of each node is minimized to some extent. The proposed algorithm can be extended to 3-dimensions. All these works on topology control to construct fault-tolerant networks by adjusting transmit power of nodes. Movement of nodes is not a controllable parameter even in the works where mobile networks are considered.

Significant amount of work has been done in coordinating teams of mobile robots or actors. However, little attention was paid to incorporate fault tolerance into these robotic networks. For example, Dynia et al. [5] studied the problem of maintaining communication between an explorer robot and base station by moving other robots along the path.

Mobile robot network can be represented as a graph, where each node is a mobile robot and each edge denotes a communication link between a pair of robots. In a connected graph, a node is called a **critical node** if the graph is disconnected without the node. There are no critical nodes in a bi-connected graph. So, critical nodes are important in designing movement control algorithms to achieve bi-connected networks. Jorgic et al. [6] proposed an approach for localized p-hop critical node detection. To find if a node is critical in the network, a sub-graph of p-hop neighbors of the node is considered. From this sub-graph, the node itself and all its incident edges are excluded. If this resulting sub-graph of p-hop neighbors of a node is disconnected by excluding the node, then the node is critical. Since only local topological information is used, it is specified as p-hop critical node as it may not be globally critical. However, all the globally critical nodes are always p-hop critical for any value of p. As seen in Figure 1, the nodes A, B, and C are 2-hop critical nodes in the given network. We can also notice that nodes A and B in Figure 1 are only 2-hop critical nodes and are not globally critical. However, node C is globally critical in the network and is also 2-hop critical. Experiments showed that over 80% of locally estimated critical nodes and links are indeed globally critical [6].

Our problem is most related to the problem discussed by Basu and Redi [2], where movement control algorithms for fault-tolerant robot networks were proposed. In these algorithms, each mobile robot was assumed to be aware of global network topology. Based on the topological information, robots decide on their new position, which would thereby create a fault-tolerant network. The goal of these algorithms was to minimize the total distance travelled by all the robots. The authors further proposed an approximation algorithm for two dimensional cases. The basic idea was to divide a network into bi-connected blocks. The network is a block tree of these blocks. A block with maximum number of robots acts as the root of the tree. Algorithm works iteratively merging the blocks to form a single bi-connected block. Merging of the blocks is performed by block movement where each leaf block is moved towards its parent. If parent

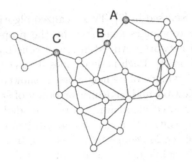

Fig. 1. An example of a network containing critical nodes.

block is empty then leaf block is moved towards a critical node. After each iteration, robot connectivity is recalculated and block tree is reconstructed as well. However, the proposed algorithms require accurate and global information of entire network. It is applicable to only small size networks. For large scale networks, not only is global network information hard to obtain and maintain, but also the total distance of movements and the communication overhead on robots increase rapidly.

3   Localized Movement Control

In this section, we propose a localized movement control algorithm for fault tolerance of mobile robot networks. To the best of our knowledge, it is the first localized movement control algorithm to achieve bi-connected network topologies. For simplicity, we use a node to denote a mobile robot for the rest of paper. We assume that all nodes in the network have a common communication range r. We further assume that each node has information of its p-hop neighbors. It is can be achieved by exchanging or relaying HELLO messages periodically within p-hops. To reduce exchange packets and collisions, we assume there is no RTS / CTS mechanism for transmissions of control packets. The network is assumed to be connected but not bi-connected. The problem of our concern is to control movement of nodes, such that the network becomes bi-connected. The objective is to minimize the total distance moved.

The distributed algorithm is executed at each node and starts as follows. At initialization stage, each node checks whether it is a p-hop critical node [6]. We define the p-**hop sub-graph** of a node by the graph which contains all nodes that are within p-hops from the node and all corresponding links. A node is said to be a p-**hop critical node** if and only if its p-hop sub-graph is disconnected without the node. Since each node is assumed to have knowledge of its p-hop sub-graph, it is able to determine whether it is a p-hop critical node. If a node finds itself a p-hop critical node, it broadcasts a *critical announcement* packet to all its direct neighbors.

To make the network bi-connected, all critical nodes should become non-critical by movement of nodes. Note that the movement of a node may create new neighbors, but it may also break some existing links. Since a critical node is the node that leaves its p-hop sub-graph disconnected without itself, breaking some current links of a critical node may cause disconnection of the network. However, for a non-critical node, the network remains connected if one of its current links is broken. Our basic idea of movement control is to move non-critical nodes while keep critical nodes static unless they become non-critical. According to number of critical neighbors of a critical node, there are three different cases that need to be considered. We start from the simplest case and discuss the three cases one by one.

3.1 Critical node without critical neighbors

In this case, a node finds itself a p-hop critical node and does not receive any *critical announcement* packet from its neighbors. Since it is a critical node, its p-hop sub-graph can be divided into two disjointed sets without the node. The basic idea is to select two neighbors from two sets respectively and move them towards each other until they become neighbors. Suppose distance between the two neighbors is d. Each node should move $(d - r)/2$ to reach each other. To minimize the total distance of movement of nodes, two neighbors with the minimum distance d among all possible pairs in the two sets are selected. The critical node sends these two neighbors a *movement control* packet containing their new locations. The two neighbors move to their new locations once the *movement control* packet is received. Note that a non-critical node may have several critical neighbors and it may receive multiple *movement control* packets from different critical nodes. Node IDs are used to assign priorities to critical nodes. Therefore, if a non-critical node receives more than one movement control packets, it always follows direction by the critical node having the largest ID. Note that there is no RTS/CTS mechanism in the network. The critical nodes with smaller IDs do not know and have no need to track the movement of their non-critical neighbors after sending *movement control* packets.

After movement of nodes, any node that loses a current neighbor or finds a new neighbor broadcasts a topology updated packet to its neighbors. This packet will be relayed hop by hop to reach p-hops neighbors of the sender. Each node receiving the topology updated packet updates its p-hop sub-graph and checks its new status. A new iteration of movement control begins. The movement control algorithm for case 3.1 is illustrated with the following example.

Consider the example shown in Figure 2, where node 3 in grey color is critical node and node 1, 2, 4, 5, 6, 7, 8 in white color are non-critical nodes. Suppose $p = 2$ in this example. Since node 3 is critical, its 2-hop sub-graph is divided into two disjointed sets $A = \{1, 2, 4, 5\}$ and $B = \{6, 7, 8\}$. Suppose distance of node 5 and 8 is the minimum among all possible pairs in these two sets, i.e. $d(5, 8) \leq d(x, y), \forall x \in A, y \in B$. Node 3 computes new locations of node 5 and node 8 and sends *movement control* packets to them. Final locations of node 5 and node 8 are shown in Figure 2.

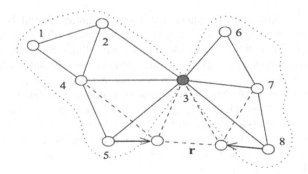

Fig. 2. Critical node without critical neighbor.

3.2   Critical node with one critical neighbor

In this case, there are two adjacent critical nodes and each critical node has only one critical neighbor. Suppose the two adjacent critical nodes are node 4 and node 5, and ID of node 5 is larger than ID of node 4. Our basic idea is to let the critical node with larger ID, node 5 in this case, select one of its non-critical neighbors to move towards the other critical node, node 4. Similar to case 3.1, node 5 divides its p-hop sub-graph into two disjointed sets. Node 4 is sure to be contained in one of the two sets. Node 5 searches the other set and selects one of its non-critical neighbors that is the nearest to node 4. Suppose distance between the selected neighbor and node 4 is d. The selected neighbor should move distance $d - r$ to reach node 4 since critical nodes are not allowed to move, to avoid disconnection of networks. Node 5 computes new location of its moving neighbor and sends it a *movement control* packet. The neighbor moves to its new location after receiving the *movement control* packet. Similar to case 3.1, node ID's are used to break the tie when a non-critical node receives multiple movement control packets. Once one of critical nodes becomes non-critical, we return to case 3.1 and movement control algorithm for case 3.1 is executed. Similar to case 3.1, topology update and checking status operations will start after movement of nodes. The algorithm for case 3.2 is illustrated with the following example.

Consider the example in Figure 3, where nodes 4 and 5 in grey color are critical nodes and nodes 1, 2, 3, 6, 7, 8 in white color are non-critical nodes. Since ID of node 5 is larger than ID of node 4, node 5 leads movement control. Suppose $p = 2$ again. Node 5 divides its 2-hop sub-graph into two disjoint sets $A = \{1, 2, 3, 4, 6\}$ and $B = \{7, 8\}$. Suppose distance of node 4 and 7 is the minimum among all neighbors in B. That is, $d(4,7) \leq d(4,x)$, $\forall x \in B$. Node 5 computes new location of node 7, and sends it a *movement control* packet. Final location of node 7 is shown in figure 3.

Note that we cannot simply select one of node 5's neighbors that is the nearest to node 4 to move. It is because there may already exist links between node 4 and node 5's neighbors. For example in Figure 3, node 3 is the nearest node to

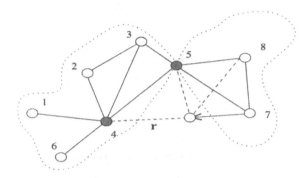

Fig. 3. critical node with one critical neighbor.

node 4 among all node 5's neighbors. However, they are already connected and there is no benefit of moving node 3 towards node 4.

3.3   Critical node with several critical neighbors

In this case, some critical node has more than one critical neighbor. Note that each node sends a *critical announcement* packet to all its direct neighbors if it finds itself a p-hop critical node. After that, all nodes in the network know the status of their neighbors. We say a critical node is **available** if it has non-critical neighbors and is **non-available** otherwise. A critical node is available means that it has non-critical neighbors that are able to move. An available/non-available critical node broadcasts an *available/non-available announcement* packet to its neighbors. A critical node declares itself a **critical head** if and only if it is available and its ID is larger than the ID of any available critical neighbor, or has no available critical neighbors. Our basic idea for general cases is to use the pair wise merging strategy. Simulation results show that this strategy can efficiently and quickly construct a bi-connected network. Each critical head selects one of its critical neighbors to pair with. Any criterion for selecting will work. To be deterministic, we decide that available critical neighbor (if any) with largest ID is selected, or otherwise non-available critical neighbor with the largest ID. Then the movement control algorithm for case 3.2 is called for each pair to compute the new topology.

Consider the example in figure 4, nodes 1, 2, 3, 4, 5, 6 in grey color are critical nodes (dashed block with a node is sub-graph of this node). Among these critical nodes, only nodes 1, 5, 6 are critical heads. Node 1 becomes a critical head since node 3 is non-available. Finally, there are three pairs: (1,3), (5,4) and (6,4), dominated by nodes 1, 5, and 6, respectively. Each critical head in a pair calls the movement control algorithm for case 3.2 to merge the pair. One can expect that the network density would increase after merging. Pair-wise merging continues until all critical nodes become non-critical, i.e., the network is bi-connected.

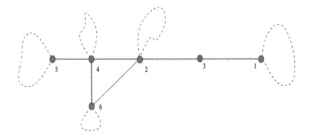

Fig. 4. Critical node with several critical neighbors.

Note that a critical head dominates a pair to merge at each time. No action will be taken if there is no critical head in the network. So the remaining problem is: does there always exist critical heads if the network is connected but not bi-connected? We answer this question below:

Lemma 1. *There exists non-critical nodes in any connected network.*

The Proof is trivial and has been omitted.

Theorem 1. *If the network is connected but not bi-connected then it has a critical node without critical neighbor or a critical head.*

Proof. According to Lemma 1, there exist non-critical and critical nodes in the network. Since the network is connected, some critical nodes will be then connected to some non-critical nodes. That is, some critical nodes are available. Consider among them the one with the largest ID. This node either has no critical neighbor or is a critical head by the definition.

According to Theorem 1, there always exist critical heads in the network if it is still not bi-connected. So, each node can iteratively call our algorithm until the network becomes bi-connected.

Note that we move only non-critical nodes in our algorithm. Controlled movement of a single non-critical node will never cause disconnection of networks. However, the network may be disconnected when multiple non-critical nodes move concurrently. It is because several non-critical nodes may form a *cut* of the network. The network becomes disconnected if critical links of nodes in the cut break. Unfortunately, the nodes of a cut can be distributed everywhere in the network. It is hardly possible for any localized algorithm to locally detect them. However, in our simulations, the proposed localized algorithm achieved 100% success on construction of bi-connected network topologies.

4   Performance Analysis

We tested the performance of our algorithm in a simulated environment and analyzed its efficiency with respect to the distance traveled metric. We also performed comparisons with the existing algorithm that uses global information [2] henceforth called the globalized algorithm. In all our simulations, the proposed algorithm was 100% successful, achieving bi-connectivity of the network within a few iterations, in most cases.

4.1   Simulation Environment

The main objective of our algorithm is to minimize the distance traveled by the robots in achieving bi-connectivity of the network. Thus, in our simulations we consider only the distance traveled metric and do not bother about the communication cost. The problem of minimizing the communication cost too, is part of the ongoing research conducted by our group. The results presented in this paper are based on simulations performed at the application layer, assuming an ideal MAC layer underneath, with no communication loss and instantaneous delivery of messages.

As for the simulation parameters, we considered sensor fields of sizes ranging from 20m X 20m to 100m X 100m, and the communication range of all the nodes were set to 10m. Nodes were placed randomly within the sensor fields, maintaining an average density of 1 node per $30m^2$ which ensures an average of around 10 neighbors per sensor node. We performed experiments varying the number of nodes in the sensor field while scaling the sensor field size accordingly. From the randomly generated networks, we selected the ones which were connected but not bi-connected and this set of networks were used in our experiments. We assume that each mobile robot has initial knowledge of its own position and can obtain information from its p-hop neighbors. We also assume that a robot can freely move from one position to another within the sensor field (i.e. there are no physical obstacles in the sensor field).

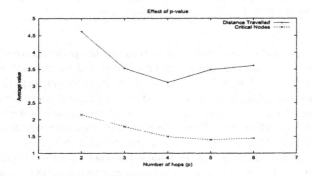

Fig. 5. Effect of the value of p on the distance traveled and the critical nodes identified by our algorithm.

4.2   Simulation Results

The p-value used in our algorithm is an important parameter that can be used to fine-tune the performance of our algorithm. Notice that if we use a small value of p, some nodes which are globally non-critical may be identified as critical by our localized algorithm. This will increase the value of distance traveled as more nodes need to be moved. We performed experiments with different values of p on a network of size 100, and the results are shown in Figure 5. As expected, both the total distance traveled and the critical nodes identified decrease with the increase in p-value. However, there is a sharp difference between $p = 2$ and $p > 2$, whereas the difference between $p = 3$ and $p > 3$ is not that significant. Since it is advisable to restrict the communication to smaller neighborhoods, we choose the value of $p = 3$, to balance the communication overhead and the distance traveled metric. The rest of the experiments were performed with the value of p set to 3.

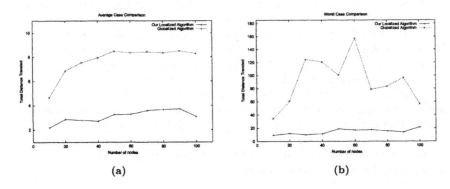

(a)   (b)

Fig. 6. Comparison of the algorithms in networks of different sizes with fixed density and average degree 10. (a) Average case (b) Worst case

We performed extensive simulations to compare the performance of our algorithm with that of the globalized algorithm. The size of the network was varied from $n = 10$ to $n = 100$, and in each case, we repeated our experiments 100 times for both the algorithms. Figures 6 (a) and (b) show the average and worst case behavior of the two algorithms, in terms of the distance traveled metric. Our proposed algorithm outperforms the globalized algorithm significantly in all cases. In our algorithm, in each iteration, individual nodes are moved towards each other instead of moving blocks of nodes together as in the case of the globalized algorithm. This results in great performance improvement as can be seen from the simulation results. Notice that, since our algorithm uses only local information to identify the critical nodes, it would identify more nodes as critical nodes as compared to the globalized algorithm. Figure 7 compares the number of nodes initially identified as critical by the two algorithms, in the average case.

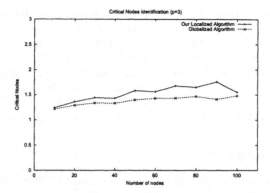

Fig. 7. Critical node identification by the two algorithms in networks of various sizes

It seems that our localized algorithm did not identify too many globally non-critical nodes as critical, which suggests that using only local information is not too harmful in general, for identifying critical nodes.

5   Conclusions and Future Work

In this paper, we proposed a localized movement control algorithm to construct a fault-tolerant mobile robot network. We presented simulations results to show the effectiveness of our algorithm and its efficiency in terms of the total distance travelled by the robots. The simulations results for randomly generated networks show that our localized movement control algorithm significantly outperforms its globalized counterpart. It is interesting to note that, in most cases, the use of local information (in fact, information about 3-hop neighbors only) is sufficient to convert the network to a bi-connected one in an efficient manner. Thus, global information about the network is not necessary to achieve bi-connectivity. The results shown in this paper are for networks obtained by randomly scattering robots on fixed region and then selecting those which satisfy the condition of connectivity and non-bi-connectivity. On this class of graphs, the proposed algorithm seems to work quite well, in fact achieving 100% success. However, it is quite possible that for some specific class of networks this algorithm may not perform that well. In particular, we tested only networks with average number of neighbors (density) about 10, and results may differ for lower densities. In future, we plan to do more extensive simulations on specially created network topologies in order to identify possible classes of networks for which our algorithm may give poor performance. We are also investigating on applications of mobile robots in sensor networks as data collectors. We recognize that mobile robots can also be used in heterogeneous sensor networks as data collectors. In these applications, sensor field coverage is an important metric for the algorithm and would be the topic for future work.

The localized algorithm proposed in this paper achieves fault-tolerance by converting a connected network to bi-connected one. In case the original network is disconnected, the algorithm can be used to make each connected component fault-tolerant. However, the problem of constructing a connected and fault-tolerant network starting from a disconnected network is much more difficult and would be considered in future. Finally, we did not consider the problem of minimizing communication cost as added criterion, although it was addressed by localized design. That problem is currently under investigation by this group.

Acknowledgements

This research is supported by NSERC Strategic Grant STPGP 336406, and the UK Royal Society Wolfson Research Merit Award.

References

1. M. Bahramgiri, M. Hajiaghayi, and V. Mirrokni. "Fault-tolerant and 3-dimensional distributed topology control algorithms in wireless multi-hop networks", In Proc. IEEE Int. Conf. on Computer Communications and Networks (ICCCN02), pp. 392–397, 2002.
2. P. Basu and J. Redi, "Movement control algorithms for realization of fault-tolerant ad hoc robot networks" , IEEE Network, 18(4):36-44, 2004.
3. R. Chow, and T. Johnson, "Distributed Operating Systems & Algorithms" , Addison Wesley Longman Inc., 1997.
4. J. Cheriyan, S. Vempala, and A. Vetta, "An Approximation Algorithm for the Minimum-Cost k-Vertex Connected Subgraph" , SIAM J. Computing 32(4): 1050-1055, 2003.
5. M. Dynia, J. Kutylowski, P. Lorek, and F.M. auf der Heide, "Maintaining Communication Between an Explorer and a Base Station" , in IFIP Conf. on Biologically Inspired Cooperative Computing (BICC'06), vol. 216, pp. 137-146, 2006.
6. M. Jorgic, I. Stojmenovic, M. Hauspie, D. Simplot-Ryl, "Localized Algorithms for Detection of Critical Nodes and Links for Connectivity in Ad Hoc Networks " , In 3rd Ann. Med. Ad-Hoc Workshop (Med-Hoc-Net), pp. 360-371, 2004.
7. N. Li, and J.C. Hou, "Topology Control in Heterogeneous Wireless Networks: Problems and Solutions" , In Proc. IEEE INFOCOM, pp. 232-243, 2004.
8. N. Li, and J.C. Hou, "FLSS: A Fault-Tolerant Topology Control Algorithm for Wireless Networks", in Proc. 10th Ann. Int. Conf. on Mobile Computing and Networking, pp. 275-286, 2004.
9. H. Liu, P.J. Wan, X. Jia, "Fault-Tolerant Relay Node Placement in Wireless Sensor Networks" , in Proc. COCOON, pp. 230-239, 2005.
10. R. Ramanathan and R. Rosales-Hain, "Topology Control of Multihop Wireless Networks using Transmit Power Adjustment", In Proc. IEEE Infocom, vol.2, pp. 404-413, 2000.
11. D. Vass, A. Vidcs, "Positioning Mobile Base Station to Prolong Wireless Sensor Network Lifetime" , in CoNEXT 2005 Student Workshop, pp. 300-301, 2005.
12. O. Younis, S. Fahmy, and P. Santi, "Robust Communications for Sensor Networks in Hostile Environments", In Proc. 12th IEEE Int. Workshop on Quality of Service (IWQoS), pp. 10-19, 2004.

Intelligent Actor Mobility in Wireless Sensor and Actor Networks

Sita S. Krishnakumar and Randal T. Abler

Georgia Institute of Technology, Savannah GA 31407, USA
{sita,randal.abler}@gatech.edu

Abstract. In wireless sensor and actor network research, the commonly used mobility models for a mobile actor are random walk model, random waypoint mobility model, or variants thereof. For a fully connected network, the choice of mobility model for the actor is not critical because, there is at least one assured path from the sensor nodes to the actor node. But, for a sparsely connected network where information cannot propagate beyond a cluster, random movement of the actor may not be the best choice to maximize event detection and subsequent action. This paper presents encouraging preliminary results with static intelligent mobility models that are found using the inherent clusters' information of a sparsely connected network. Finally, a proposal to develop dynamic intelligent mobility models for the actor based on a mathematical model is presented.

Key words: mobile actor, mobility model, intelligent movement, clusters

1 Introduction

The concept of a mobile actor is relatively new in the area of wireless sensor and actor networks (WSAN). The main advantage of having a mobile actor is that more effective action can be taken, since the actor can get as close as possible to the event location. Additionally, in a densely connected network the presence of a mobile actor eases the problem of overburdened forwarding nodes by balancing the load among all the nodes in the network. In a sparsely connected network, a mobile actor can bridge the connectivity between groups of isolated nodes. But, there is no guarantee that the existing mobility models will work efficiently for a mobile actor in a sparsely connected network. A sparsely connected network can be used in applications where < 100% event detection is acceptable. An example application is coastal monitoring where sensor nodes may be deployed to monitor water temperature, wave characteristics, water level or meteorological conditions. When an event occurs in this environment, a possible action taken by the actor node could be of collecting a water sample for further analysis.

In a sparsely connected network, there are naturally occurring groups of nodes within which information can be sent. These node groupings are referred to as *clusters*. An event can be detected by one or more sensor nodes belonging to one or more clusters. The event detection information can then be propagated throughout the cluster which has detected the event, but cannot be sent beyond

---

Please use the following format when citing this chapter:

Krishnakumar, S. S., Abler, R. T., 2007, in IFIP International Federation for Information Processing, Volume 248, Wireless Sensor and Actor Networks, eds. L. Orozco-Barbosa, Olivares, T., Casado, R., Bermudez, A., (Boston: Springer), pp.13-22.

that cluster. The information can reach the actor only if the actor visited the cluster which has detected the event. It would be valuable if the mobile actor followed a path such that the number of events detected in a timely manner can be maximized. This paper presents a novel way to generate static intelligent mobility models for the actor, using the inherent clusters' information of a sparsely connected network. The rest of the paper is organized as follows: Section 2 discusses related work with mobile actors and Section 3 presents the methodology used to develop the static mobility models. Section 4 presents the simulation results and Section 5 discusses the future work to develop dynamic intelligent mobility models. Section 6 summarizes and concludes the paper.

2   Related Work

The commonly used mobility models in WSAN research are the random walk mobility model, random waypoint mobility model, or a variation of the two. In the case of the random walk model, the actor moves towards a random destination at a certain speed. On reaching the destination, it chooses yet another random destination and continues towards that destination, and so on. In the case of the random waypoint mobility model, random destinations are picked uniformly in the region and the actor moves at a selected speed which is also chosen uniformly in an interval. Upon reaching the destination, the actor pauses for a time period, and the process repeats itself afterwards. Random mobility models are non-intelligent mobility models. When evaluating the performance of algorithms and protocols, they are suitable because the results obtained are then not tailored to specific mobility models. When using random mobility models, evaluating against a fully connected network assures that there is always a path available between the sensor nodes and the actor nodes for information exchange.

One of the prime requirements of a WSAN is that events have to be detected in a timely manner and communicated to the actor at the earliest possible time [1]. Since a centralized sink may not be present in a WSAN, coordination and communication between the actors and between the sensors and the actors have to be managed. The coordination and communication problem has been studied and a hybrid location management scheme has been proposed to handle the mobility of actors in [8]. This approach deals with a fully connected network and the actor moves following a variation of the random waypoint mobility model. Though the action and movement energy of the actor are assumed to be orders of magnitude higher than the energy required for communication purposes, this paper does not analyze the impact of random actor movement on the movement energy.

Another research paper that deals with mobile actors proposes a real-time coordination and routing framework for sensor-actor coordination to achieve energy-efficient and reliable communication [9]. The sensor nodes in the network form clusters based on application parameters and elect a cluster-head, which is responsible for communicating with the actor node. A backbone network of forward-tracking and backtracking mechanisms is formed, which helps in routing the packets. The main goal of the framework is to deliver event detection information to the actor within a given delay bound. When the actor is mobile, the deadlines miss-ratio does not reach zero even with sufficient delay. This could either be due to congestion in the network or due to the random movement of the actor. These results highlight the fact that when routing is done hierarchically using only cluster-heads, random movement of the actor may not be the best choice for data collection.

Mobile entities have also been used in wireless sensor network (WSN) and mobile ad-hoc network (MANET) research. Data mules are mobile entities that collect data from sensors as they pass by [10]. They were introduced to help with non-real-time data collection in sparsely connected sensor networks. In simulations, the mules followed a random path or a deterministic path [4]. Mobile elements were also used for data collection in sparsely connected networks [3]. Nodes were partitioned into bins based on geographical locations and buffer overflow times. The schedule for each bin was created by solving the traveling salesman problem. These individual paths were concatenated to form the overall path of the mobile element. Message ferrying is a proactive routing scheme that was introduced in highly partitioned wireless ad-hoc networks where network connectivity can be maintained only when the nodes use their long range transmission radio [11]. The message ferry basically reduces all communication to a single hop at the cost of delay. The route taken by the ferry is found using a traveling salesman problem approximation followed by optimization of average delay.

The mobile entities deployed in WSNs and MANETs were meant to collect data from every single node in the network. In our research, the sparsely connected networks have naturally occurring node groupings within which the nodes can communicate with each other. When an event occurs in a cluster, all the nodes in the cluster will receive the event detection information using propagation techniques. This paper uses the clustering information of the network to help define a path for the mobile actor, with the goal of maximizing the number of events detected in a timely manner. To the best of our knowledge, there is no known research work in the WSAN area where the mobility model of the actor is reflective of the network in which the actor is deployed.

3   Methodology

To devise a mobility model for the actor in a sparsely connected network, the clusters of a given network has to be found. Pseudo code to generate the clusters' information from the sensor node locations and the sensor node transmission range is shown in Fig. 1. Sensor node locations can be obtained using localization techniques such as angle-of-arrival measurements, distance-related measurements, or received signal strength profiling techniques [6].

Based on the clusters' information, a decision has to be made as to which clusters the actor will visit and when. A large number of sensor nodes in a cluster

```
/* find the neighbors of every node in the network */
/* tx_range - transmission range (diameter) of a sensor node */
for(every node i in the network ) {
    for(every node j in the network ) {
        if( node i != node j ) {
            if( distance between node i and node j < tx_range ) {
                node i and node j are neighbors;
            }
        }
    }
}

/* find paths emanating from every node */
for(every node i in the network ) {
    find all possible paths in the network;
    cnt[i] = unique number of nodes reachable from node i;
}

/* cnt[i] denotes the size of the cluster to which node i belongs */
for(every node i that has not been assigned to a cluster ) {
    k = number of nodes that have the same cluster size as node i;
    if( k == cnt[i] ) {
        /*all k nodes are part of the same cluster*/
        assign a cluster_id and associate the nodes;
    }
    if( k > cnt[i] ) {
        go through the paths and find clusters of
                size cnt[i] from the k nodes;
        assign each cluster a cluster_id and associate
                the corresponding nodes with that cluster;
    }
    mark the k nodes as having been assigned to a cluster;
}
```

Fig. 1. Pseudo code to find the clusters in a network

does not automatically qualify the cluster for a visit. The parameter of interest is the area occupied by the cluster. This is because, the larger the area of a cluster, the greater the chance of an event occurring in that cluster. The clusters' area can be found using the pseudo code shown in Fig. 2.

```
for( every square meter of the field ) {
    for( every node n in every cluster m ) {
        if( distance between node n and center of this sq. m. < tx_range/2 ) {
            mark this sq. m. as covered by node n of cluster m;
            increment cluster m's area by 1 sq. m.;
        }
    }
}
```

Fig. 2. Pseudo code to find the area of clusters in a network

Given that the probability of an event having occurred is $P_E(t)$, the probability of an event having occurred in a cluster $P_{E_c}(t)$ is given by

$$P_{E_c}(t) = \frac{A_c}{A_t} * P_E(t) . \tag{1}$$

where A_c is the area of the cluster and A_t is the total area. The clusters with a higher probability of an event occurrence are those that occupy more area and when visited by the actor may lead to better event detection results.

A mobility model generates the path taken by a mobile actor and is defined by

1. A set of points S and
2. An order in which to visit the points in S to generate a path P.

When an event occurs in a cluster, the event detection information is propagated throughout the cluster. All the nodes in a cluster will have the same information over a period of time. The time required for this information to be propagated throughout the cluster will vary depending on the location of the event and the area of the cluster. In this preliminary research, the centers of the clusters are used as points to visit. The center of a cluster can be found by averaging the x and y coordinates, respectively, of all the sensor nodes belonging to the cluster. This ignores the possibility that the geographic center of the cluster may not be part of the cluster. For those clusters where the probability of an event occurrence is high, the centers of the clusters are found and added to S. Now, the points in S have to be ordered to find the path P. Any point in S can be chosen as a starting point. The following approaches are just a couple of ways to find P:

1. Using the *nearest-neighbor* approach: This approach follows a greedy algorithm similar to a solution for the traveling salesman problem. Fig. 3 shows the steps involved in generating a path using this approach. Depending on the speed of the actor, the duration of a simulation may be longer than the time it takes to visit the points in S once. If this is the case, then after all the points in S have been visited once, the last point becomes the new starting point and Steps 2-4 are repeated to generate the path P for the required duration.

2. Using the *nearest-x-coordinate* approach: In this approach, a fixed list denoted by L is generated, as shown in Fig. 4. For a starting point s, scan the list L to find s. Add the point to P. The list L is then browsed forward and points are added to P. When the last point in L is reached, L is browsed backwards and points are added to P. When the first point in L is reached, L is browsed forward again and so on for the duration of the simulation. With this approach, the field is traversed in a zigzag manner along the y-axis. Some points may be visited more than once before all the points in S have been visited once. Similarly the nearest-y-coordinate approach can also be used where the field is traversed in a zigzag manner along the x-axis. In this work, only the *nearest x-coordinate* approach is used in experiments.

---

1. Add the starting point to P and mark it as visited in S.
2. Select the closest unvisited point to visit next. Add that point to P and mark it as visited in S.
3. Repeat Step 2 until there are no more unvisited points in S.
4. Mark all points in S as unvisited.

Fig. 3. Steps to generate a path using the nearest-neighbor approach

---

1. Find the point in S with the smallest x-coordinate value. Add this point to L and mark it as visited in S.
2. Select an unvisited point in S with the closest x-coordinate to the current point to visit next. Add that point to L and mark it as visited in S.
3. Repeat Step 2 until there are no more unvisited points in S.
4. Mark all points in S as unvisited.

Fig. 4. Steps to generate an ordered list using the nearest-x-coordinate approach

4   Preliminary Results

4.1   Simulation Environment

The simulations were run against a network of 100 sensor nodes in a 200 m x 200 m field with one actor node. Sensor node locations were generated using a uniform random number generator. The routing protocol used was an optimized broadcast protocol known as BCAST [5]. This protocol keeps track of one-hop and two-hop neighbors and sends information only to them. Only packets that would reach additional neighbors are re-broadcast. Other simulation parameters are shown in Table 1. The stationary, fixed duration events are uniformly distributed in space and time. The terminating simulations were carried out using the network simulator ns-2 [7]. The events were represented in ns-2 using extensions from the Naval Research Laboratory [2].

Table 1. Simulation parameters

Number of sensor nodes	100
Number of actor nodes	1
Transmission range: Sensor node	20 m
Transmission range: Actor node	30 m
Actor speed	2 m/s
Field dimension	200 m x 200 m
Event diameter	30 m
Event duration	100 seconds
Number of events	10

For the example network, the clusters' information and their area were generated using the steps shown in Figs. 1 and 2 respectively. There were 23 clusters ranging in size from 1 to 24. The largest cluster occupied 11.86% of the total area, while the smallest cluster occupied 0.39% of the total area. The probability of an event having occurred in a cluster was found for all the clusters. For the example network, the clusters with a higher number of nodes corresponded to a larger area and hence a higher probability of an event occurrence. For the simulations, 14 clusters where the cluster size was greater than 1 were selected. The chosen clusters' centers were then calculated to be used as points to visit.

The simulation duration was 400 seconds. Actor paths were generated for the simulation duration using the two approaches defined in Section 3. An actor path following the random walk mobility model was generated using tools available in ns-2 for 400 seconds. The actor speed for random movement was also set at 2 m/s.

Table 2. Simulation results (*italic* entries highlight values obtained using intelligent mobility models - NN: nearest-neighbor approach, NX: nearest-x-coordinate approach - that are better than values obtained using RW, the random walk model)

	Average time (secs)			Worse time (secs)			Events detected		
	NN	NX	RW	NN	NX	RW	NN	NX	RW
Run 1	*23.61*	*13.09*	42.64	86.00	*40.98*	79.95	*7*	*5*	3
Run 2	*0.04*	*18.32*	28.08	*0.11*	*50.14*	56.15	*4*	*4*	2
Run 3	59.74	*4.09*	18.35	100.11	*20.12*	35.89	*5*	*5*	3
Run 4	*22.76*	*14.69*	25.30	68.01	*47.95*	56.25	3	*5*	4

4.2   Simulation Results

For each of the three mobility models, four simulations were run. The starting point of the actor was varied in each run. For each run, the time to detect every event was noted. From this, the average time to detect an event was found. This, along with the worse time to detect an event and number of events detected are tabulated, as shown in Table 2.

The average time to detect an event with the proposed intelligent mobility models fared better than the random walk mobility model in most of the individual runs. The worse time to detect an event was not as conclusive. This is because the mobility models used in the simulations were preliminary in design with no optimization performed. The proposed mobility models also detected more events in most runs than the random walk model, as shown in Table 2. The clear winner is mobility model *NX* - the mobility model going zigzag along the y-axis using the nearest-x-coordinate approach with center point of clusters - and it outperformed the random walk model in every run.

5   Future Work

The preliminary results validate the assumption that a random walk model is not the best mobility model for the actor in a sparsely connected network. The intelligent mobility models that have been defined in this paper are static in nature and do not take into account the time an actor may spend acting. The paths generated by these static mobility models can be calculated in advance and can be used in situations where the actor's movement has to be pre-loaded. The future goal is to optimize the actor's path in a dynamic environment where, when an event occurs, the actor may be involved in acting for an unspecified duration of time. In order to be able to incorporate the dynamic acting pattern of the actor, we plan to define a mathematical model for fixed duration events. This mathematical model would be implemented at the actor node and would help in defining the expected number of detectable events in a cluster and the expected number of lost events (events that occurred but went undetected) in a cluster. Depending upon an application's requirements, different dynamic mobility models will be generated for the actor node using the mathematical model.

The results of the dynamic mobility models will then be compared against the static mobility models that have been presented in this paper.

With a sparsely connected network and sparse number of actors, approaching 100% detection may not be possible. We plan to conduct extensive simulations to study how close to 100% detection we can reach with optimization.

6 Summary

Static intelligent mobility models for the actor were generated using the clusters' information of a sparsely connected network. Even though random walk is not an intelligent mobility model, we used it for a baseline comparison because of it's prevalence in literature for use in both fully connected and sparsely connected networks. With one of the proposed intelligent mobility models, on an average 60% more events were detected than with a random walk model. Also, the average time to detect an event was on an average 60% lesser than the results with a random walk model. These results have validated the idea of using the inherent characteristics of a sparsely connected network to generate intelligent mobility models for the actor, which helps in detecting more events in a timely manner. A simple yet effective mathematical model will be developed, that would aid in generating dynamic intelligent mobility models for the actor based on application requirements. These dynamic mobility models will be evaluated against different application parameters and the results will be reported in a future paper.

References

1. Akyildiz, I., Kasimoglu, I.: Wireless sensor and actor networks: Research challenges. Ad Hoc Networks Journal (Elsevier) (2004) 351–367
2. Downard, I.: Simulating Sensor Networks in NS-2, NRL/FR/5522-04-10073, Naval Research Laboratory, Washington, D.C., May 2004
3. Gu, Y., Bozdag, D., Ekici, E., Ozguner, F., Lee, C-G.: Partitioning based mobile element scheduling in wireless sensor networks. Proceedings of IEEE Secon'05 (2005) 386–395
4. Jain, S., Shah, R., Brunette, W., Borriello, G., Roy. S.: Exploiting mobility for energy efficient data collection in wireless sensor networks. Mobile Networking Applications (2006) 327–339
5. Kunz, T.: Multicasting in mobile ad-hoc networks: achieving high packet delivery ratios. CASCON '03: Proceedings of the 2003 conference of the Centre for Advanced Studies on Collaborative research 156–170
6. Mao, G., Fidan, B., Anderson, B.: Wireless sensor network localization techniques. Computer Networks 2007
7. McCanne, S., Floyd, S.: ns-Network Simulator, Available from http://www-mash.cs.berkeley.edu/ns/
8. Melodia, T., Pompili, D., Akyildiz, I.: A Communication Architecture for Mobile Wireless Sensor and Actor Networks. Proceedings of IEEE Secon'06 (2006)

9. Shah, G. A., Bozyigit, M., Akan, O. B., Baykal, B.: Real-Time Coordination and Routing in Wireless Sensor and Actor Networks. 6th International Conference, NEW2AN 2006

10. Shah, R., Roy, S., Jain, S., Brunette, W.: Data MULEs: Modeling a Three-tier Architecture for Sparse Sensor Networks. IEEE SNPA Workshop, 2003

11. Zhao, W., Ammar, M.: Message Ferrying: Proactive Routing in Highly-Partitioned Wireless Ad Hoc Networks. FTDCS '03: Proceedings of the The Ninth IEEE Workshop on Future Trends of Distributed Computing Systems

Analysis techniques and models for resource optimization in Wireless Sensor/Actuator Network environment

Salvatore F. Pileggi, Carlos E. Palau, Manuel Esteve

ETSIT. Universidad Politécnica de Valencia
Camino de Vera S/N, 46022 Valencia, Spain
salpi@doctor.upv.es, {cpalau, mesteve}@dcom.upv.es

Abstract. In the last few years, WSN has been object of an intense research activity that has determined an important improvement by technologic and computation point of view both. The notable level got and the increasing request of applications designed over Sensor Networks make WSN commercial diffusion next to be a fact. Limited resource orientation and high level application requirements result in a number of key open issues, such as Resource Optimization and Quality of Service. These last two issues require an important preliminary phase of analysis and evaluation that can provide the designer with knowledge of important relationships between parameters design and application desired characteristics. Mathematical models of local resource (node), of network influence on single resource, of QoS requests, and related analysis techniques to determine not only "how much" but also "in which way" resources are expensed are proposed in this paper.

Keywords: Resource Analysis, Resource Evaluation and Optimization, Wireless Sensor Network, Quality of Service.

1 Introduction

The intense research activity of the last few years has determined an important improvement for WSN [1] by technologic and computation point of view. The last generation of sensor/actuator nodes, even if limited resource oriented in terms of power, is provided with ad-hoc miniaturization of advanced micro-processors, high level memory of important size, high level and low power communication possibilities based on several standards, very sophisticated sensor boards and high level operating systems. The technologic level of basic nodes reflects an important global improvement for systems designed on WSN that determines great research and commercial interest. Application level and commercial requirements address a number of key issues (and related challenges) that can globally referred as Quality of Service (QoS) requirements [2]. Some of these issues (security, reliability) both with some aspects related to compatibility could represent the real key issues for commercial diffusion of WSN in the next future. The dominant theme that characterizes the deployment of application on WSN is the optimal middleware between increasing application requests and (limited) resource optimization. Architectures, related topologies and associated mechanisms are always more

Please use the following format when citing this chapter:

Pileggi, S. F., Palau, C. E., Esteve, M., 2007, in IFIP International Federation for Information Processing, Volume 248, Wireless Sensor and Actor Networks, eds. L. Orozco-Barbosa, Olivares, T., Casado, R., Bermudez, A., (Boston: Springer), pp. 23-34.

complex and, in the great part of cases, imply a number of tradeoffs that could be in contrast between them (scalable/reliable routing or high performance routing? e.g.).As consequence, classic analysis techniques for evaluation and optimization of resource, even if not obsolete, can be considered, if related to high complexity of networks, as not suitable and, some times, not precise. These aspects advise to re-consider the term "resource optimization", generally referred, in WSN environment, "simply" as energy consumption minimization. This last concept remains generally valid but it has to be particularized into a great set of environments, requirements and characteristics completely different between them (hard real time and non real time, for example). Classic approaches are oriented to answer the question "how much?". This should be considered only as the first step and integrated with other analysis steps that should be the answer for two other questions: "in which way?" and "is system optimized in according with a number of desired characteristics?". The first issue is mainly related with network design in function of QoS desired characteristics; the second point should have a comparative approach oriented to resource optimization. This work would provide a basic and extensible analysis framework for high computing WSN. The first part of the paper is focused on resource model; theorist and general assumptions are related to concrete motes [3] and to external influence in function of QoS structures and mechanisms. Some aspects, related to analysis and optimization of resource, are described and, finally, some simulation results related with three different scenarios are showed.

2 Related Work

WSN is a WPAN composed by particular computational nodes provided with sensing/actuating capacity and characterized by limited resource. Its peculiar characteristics naturally orient the research on WSN to resource management/optimization. There is a considerable number of works oriented to modify and adapt mechanisms and schemes designed for traditional Wireless Networks to limited resource environment; at the same time, there are several works focused on guarantee one or more QoS properties [6] in according with resource optimization. If level considered is referred to the last generation of WSN, probably there is not an unique model that resumes, in complete and global way, the problem of resource optimization in function of a number of QoS desired characteristics and related tradeoffs and challenges. This work would be an example of global analysis oriented to provide the designer with an important tool for evaluation and comparison between them of different and complex architectures and solutions.

3 Resource Model

The analysis proposed is first focused on single computational node, considered as active part of a number of network tasks; than, node is contextualized and external influence is related with local resource expense. Moreover, information related to single nodes provides an advanced and abstract mathematical point of view of considered platform. Analysis is supposed to be event discrete and so mathematical expressions and relationships are always related to generic event i.

3.1 Node Resource

Node model considered and referred to as Node Resource is provided with five different resources:
- Computational Resource
- Storage Resource
- Communication Resource
- Sensing Resource
- Actuating Resource

As reference, the node is supposed to be provided with the same technologic level of MICA2 family [3]. Every resource of a node really contributes in different way to power consumption. The power consumed from event zero until a generic event i is defined in (1); P_i is power consumption related with single event i; P_{max} is maximum battery capacity; k is simply a scale factor.

$$Power_k = k / P_{max} * \sum_{i=0}^{k} P_i \tag{1}$$

A node is supposed as provided of a sleep/operate protocol that works in according with the activity diagram showed in Figure 1. The "natural" state for a sensor node is "Sleep"; in this state the sensor is in very low cost mode and its power expense can be supposed next to be zero; its only activity is waiting for an external or internal event; if an event, internal or external, is generated, the sensor state becomes "Operate". This is an active state and power expense must be considered as important. When all activities related with the interest event finish, sensor state becomes "Sleep".

Figure 1: Sleep/Operate protocol activity.

Figure 11: HDC analysis; TopC density diagram.

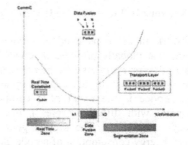

Figure 2: Transmission zones.

Active/Sleep Factor, Y, is defined in (2) and expresses the cost of Sleep/Operate protocol; Y>1 is the cost of transition Sleep→Operate. The cost of contrary transition (Operate-> Sleep) is considered to be as no influent for analysis objectives.

$$Y = \begin{cases} status == "operate" \Rightarrow Y = 0 \\ status == "sleep" \Rightarrow Y > 1 \end{cases} \quad (2) \qquad Cost_{micro-event} = f(parameters) = equation \quad (3)$$

$$Cost_{computation} = f(n - operations) = Y + k_1 * n \qquad (4)$$

A specific cost function is defined for all micro events (internal an external), in accordance with the general model of (3). Computational cost is supposed to be proportional with the number of macro operations, as defined in (4) (k_1 , scale factor). Receive cost is supposed as proportional with the percentage of information of received message (5). Transmission cost, defined in (6), is supposed proportional with percentage of information too.

$$Cost_{receive} = f(\%Info) = Y + k_2 * \%Info \qquad (5)$$
$$\%Info = SizeOf(Info) /(SizeOf(Info) + SizeOf(Head))$$

$$Cost_{transmission} = f(\%Info) = k_3 * \%Info \qquad (6)$$

When a message is received, sensor node must verify if receiver is itself (pre-process) or not. Pre-process event happens always after a receive event and, so, when sensor state is "operate" (Y=0). Pre-process cost is assumed to be constant (7).

$$Cost_{pre-process} = const \quad (7) \qquad Cost_{capture} = Y + const \quad (8) \qquad Cost_{actuate} = const \quad (9)$$

Also capture cost (8) is assumed as constant but it can happen when sensor node is in sleep or operate mode both. Actuating event is normally a consequence of a data process ad so it can happen only when sensor node status is "operate" (Y=0). Its cost is defined in (9). A node of MICA2 family [3] is considered as reference. More concretely, MICAz [3] platform is considered; MICAz has the same characteristics of MICA2 but it uses ZigBee technology [4] (based on IEEE 802.14.5 standard) that guarantee great efficiency in low power communication.

3.2 Network Influence on Node Resource

Network model considered is service oriented: the WSN is a unique virtual resource that globally provides a service; service provided is really the result of a number of tasks; every task can be represented by an activity of a unique sensor or by the activities of a group of cooperating nodes. However, and in whichever topology, global service is the result of a co-operation between all computational nodes (eventually with different "roles"), one or more base stations, and, eventually, gateway components. The section goal is to define a mathematical relationship between energy consumption of a node and external influence. This information also defines the "role" of the node into the system. External influence could really depend by a considerable number of factors, role and geographic/logic location of the node first of all. The main idea is to provide the designer of an analysis technique, based on a well-defined model that should provide a deepened knowledge of resource expense in function of a number of network parameters. Interest parameters (at network level) are considered:

- Communication efficiency: it depends mainly by application characteristics (hard/soft real-time, no real-time) and by throughput optimization.

- Topology: a node is active part of a global service and, so, of one or more computational tasks. Some of its function are active (capture, for example), others passive (transit node).
- Data Process: data process implies that node status is (or becomes) "Operate" (high energy expense). If tasks are organized in co-operative way and computation level is supposed to be high, opportune strategies are advised. However, data process importance is not directly related to power expense; data process represent an important activity "map" of the network and, if required, could provide the designer with an important knowledge related, directly or indirectly, to complex tasks and planes.
- Sensing plan: every application/architecture has its own requests in terms of information capture. Sensing is, both with communication, the main function of a node in WSN. In this sense, reliability and optimization importance advises to design (if required) capture planes that could guarantee performance, fault tolerance and reliability.
- Actuating plan: if network is also provided with actuating functions, an actuating plane should be designed in accordance with other optimization planes. This issue is more relevant respect to planes that manage the events that cause the actions (capture for example).

For each parameter, a mathematic definition is proposed. A resuming five dimensions function (*External Influence*, EI) for these issues is defined in (10); a mono dimension version of EI (linear combination of each contributes), called *MD-EI*, is also defined in (11).

$$EI_k = (\sum_{i=0}^{k} CommC_i, \sum_{i=0}^{k} TopC_i, \sum_{i=0}^{k} \Pr oC_i, \sum_{i=0}^{k} CapC_i, \sum_{i=0}^{k} ActC_i) \tag{10}$$

$$MD-EI_k = \sum_{i=0}^{k} a * b_{ki} * EI_k \tag{11}$$

$$\begin{cases} a = scaleFactor \\ b_{ki} \in [0.1] \\ k = 1,2,3,4,5 \end{cases}$$

As showed in Figure 2, in function of the percentage of information, three different zones can be distinguished: Real Time zone, Data Fusion/Aggregation zone, Segmentation zone. Real Time zone is characterized by information flows of little size and is so named because it is the typical environment of applications characterized by time constraints; the Fusion/Aggregation zone is the ideal work area by throughput point of view because the amount of information is considered to be optimal (high information percentage); apposite mechanisms (data fusion and aggregation) generally try to force applications to work constantly in this area; the third zone supposes a great amount of information that requires (or advises) data segmentation and related mechanisms (transport protocol, for example). Figure 2 shows proposed *Communication Coefficient* (CommC) too; a possible mathematical definition of CommC is proposed in (12).

$$EI_k = (\sum_{i=0}^{k} CommC_i, \sum_{i=0}^{k} TopC_i, \sum_{i=0}^{k} Pr\, oC_i, \sum_{i=0}^{k} CapC_i, \sum_{i=0}^{k} ActC_i) \tag{12}$$

$$\begin{cases} f_i \in [1,l] \\ l > 1 \\ init : l = 2 \\ TransmissionFailure: l = l + 1 \end{cases}$$

The behavior of CommC, as showed in Figure 2, is oriented to optimize throughput and energy efficiency. However, (12) has a sufficient level of dynamism, guaranteed by a number of parameters, to describe different behaviors, for example considering a different zone as desired if system specifications advise it. When a message is received, this must be pre-processed by sensor node to establish if the message receiver is itself (active role) or not (passive role). In this second case message must be re-transmitted. *Topology Coefficient (TopC)*, defined in (13), is an important evaluation of contribute to resource expense of passive activities for considered node. High values of TopC generally advise to increment network scalability because node activity is considered as too much passive.

$$TopC_i = sc * top_i * [Cost_{receive} + Cost_{pre-process}] \tag{13}$$

$$top_i = \begin{cases} node.id == message.receiver \Rightarrow top_i = 1 \\ node.id \neq message.receiver \Rightarrow top_i = w, w + + \\ init : w = 2 \\ sc = scaleFactor \end{cases}$$

Data Process Coefficient (ProC) can be defined as whichever function that has a, direct or indirect, dependence by number of elementary (or macro) operations processed or by activity time of node; an example of ProC definition is proposed in (14).

$$Pr\, oC_i = Dp * f(n) = Dp * n^2 \tag{14}$$

n=number of macro operations
Dp=scale factor

Data Process expense is no relevant in the great part of cases because its contribution to energy consumption is considered as inferior to others and, moreover, it is not always easy to measure with an appropriate accuracy level. On the other hand, ProC, as data process measurement, has great importance for analysis of activities into the network. Sensing activities are expressed by *Capture Coefficient (CapC)* defined in (15). As showed, it is supposed that a network can be provided with independent sensing tasks (no cooperative capturing) or with cooperative sensing tasks.

$$CapC_i = d * Cost_{capture} * e^{r/scc} \tag{15}$$

$$\begin{cases} IndependentSen\sin gTask \Rightarrow r = 1 \\ CooperativeSen\sin g \Rightarrow \begin{cases} init : r = 0 \\ capture : r++ \\ endOfTask : init \end{cases} \end{cases}$$

scc=scale factor
d=scale factor

Sensing task mechanisms are considered to be expensive but, at the same time, they make architectures more oriented to peer-to-peer and can guarantee accuracy in measurements, an important improvement of network robustness, an important support for failure detection and related reconfiguration operations. Considering capture planes, an important tradeoff between QoS and resource requirements is addressed. CapC could represent an important driver factor to compare different solutions. Also actuator placement and the design of action tasks could affect entire architecture design. In this sense the radius of action (measured in hops) seems as the key factor for evaluation. *Actuator Coefficient (ActC)*, defined in (16), is the expression of actuating influence (direct or indirect) on resource.

$$ActC_i = Cost_{Actuate} * e^{h/A} \tag{16}$$

A=factor scale
h=radius of action of interested actuator

As for sensing tasks, if actuator activity is considered to be important, the design of actuating planes is advised.

3.3 QoS influence on resource

Quality of Service (QoS) can be so defined: "In the fields of packet-switched networks and computer networking, the traffic engineering term Quality of Service (QoS) refers to control mechanisms that can provide different priority to different users or data flows, or guarantee a certain level of performance to a data flow in accordance with requests from the application program". Quality of Service must be considered as the real key issue when, as in WSN, resource are limited or no always sufficient if related with application requests; the importance of QoS mechanisms generally increases on large scale application fields. An approach simply "best effort" (and not regulated by QoS mechanism) in WSN could mean limited deployment strategies, higher human operations (this could be not easy and generally considered to be expensive), higher number of sensors to guarantee the same service in the same conditions and lifetime and, more generally, lower "quality" (efficiency, security, reliability, and so on). QoS is normally referred to as "QoS mechanisms"; this is because there are many parameters that can determine or affect the quality of service and so more than one protocol needs to guarantee a considerable number of desired characteristics. QoS mechanisms of a WSN are largely different to mechanisms of a traditional or standard network; they are mainly (but not only) focused on the tradeoffs between QoS and Resource Efficiency. Traditional networks are affected by classic problems as dropped packets, delay, jitter, out-of-order delivery, error, that recommend to design QoS mechanisms based on "Calling" approach or "In Advance" approach or reserving resource. Some mechanisms, or better, some of approaches proposed over traditional networks, with opportune modifications, can be applied in

WSNs too. Recently, a number of Resource Reservation mechanisms (based on soft-state agents, for example) were proposed. Probably, adapting mechanism designed for standard network to WSN is not exactly the best approach because WSN has its own design parameters (specific node hardware, specific operating systems, ad-hoc protocols and, above all, topology) and its own goals that depend mainly by application characteristics (real-time, no hard real-time, hard real-time), by system scale and by others characteristic issues. All issues (design parameters, goals) are different by the correspondent issues of a standard network. In this paper, different sub issues, more concretely security, fault tolerance, robustness, reliability and throughput are globally referred to as QoS; these issues, as well as theoretically independent, can have a direct relationship or tradeoff between them, or, as general issues, an important tradeoff with energy efficiency; resuming, all QoS parameters are indirectly related by the common and central tradeoff with energy efficiency. Function Y is an evaluation of QoS mechanisms incidence on local resource.

$$Y_i = f(QoS) \tag{17}$$

(17) should be particularized and it should explicitly represent the relationships between QoS mechanisms and node resource, modeling explicitly requirements and related tradeoffs. However, this last point, even if considered as one of central issues in resource analysis, is out of scopes of this work that is mainly focused on methodologies more than models. QoS mechanisms influence on resource is considered as a measurement of percentage of incidence on total expense.

4 Resource Analysis

Resource analysis (Figure 3) is articulated in three basic steps: modeling, measurement and evaluation/comparing/ optimization. The 3D function defined in (18) and its mono-dimensions version (19) could represent an important analysis instrument for WSN resource. Considering a 3D plane, x-axis is Power (1) and represents the real power expensed by node, y-axis is Y (17) and represents the QoS mechanisms influence on energy expense, and, finally, z-axis is MD-EI (11) and represents external influence and the role of node in the network.

$$(Power_i, Y_i, MD - EI_i) = MD - eval_i \tag{18} \qquad \sum_{i=0}^{k} (Power_i + Y_i + MD - EI_i) = eval_i \tag{19}$$

The first step is mainly related to the modeling of considered node platform on the mathematical model proposed; this implies the numeric estimation of the cost of some events/actions and the set of related scale factors; some of these estimations could result not easy or, however, not too much precise; in this last case, relationships between various events, network properties and, eventually, a number of theoretical/practices assumptions can help the designer to obtain a behavior model really next to real platform. In modeling step, optimization goal should be defined and particularized as a number of well-defined constrains or conditions over MD-EI or over some of its parameters. Modeling step is something more than a simple preliminary phase and, in some cases, could result as a hard and central step, in particular way in presence of complex relationships/tradeoffs; its affectivity generally conditions final results. Mechanisms, protocols and architectures analysis have, as general objective, to minimize the driver coefficient MD-EI (z-axis), as expression of the key idea of find an optimal middleware between desired characteristics and energy efficiency. Valid analysis perspectives are considered the x-y plane (direct relationship between total power expensed and power expensed by QoS mechanisms), x-z plane (direct relationship between power expensed and network influence) and y-z plane (direct relationship between QoS mechanisms and related activity at network level). Integral perspective (20, 21) could result interesting too.

$$\int_V (MD - eval)dv = \int_0^i (Power)dx \int_0^i (Y)dy \int_0^i (MD - EI)dz \qquad (20) \qquad \int_i (eval)di \qquad (21)$$

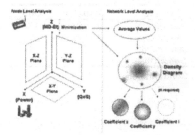

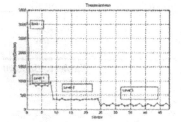

Figure 3: Resource Analysis.   **Figure 5:** Traffic Load.

Even if considering external influence, node role in the network, passive/active tasks of node in the network, the analysis limited to a single node provides a local perspective that doesn't permit an efficient evaluation/optimization step. Evaluation and optimization of resource are referred at network (or sub-network) level. Network analysis results more effective and indicative if single node measurements are extended at local area, considering values related to different nodes and mathematical/statistic elaborations of information, such as average values in certain conditions; the final result of measurement step is the definition of density diagrams of resuming evaluation functions, or, better, the definition of density diagrams related to each interest coefficient of MD-EI. Independence of optimization goal, analysis can be referred to all sensors, providing an *Area Density Diagram* (ADD) or can be restricted only at a number of key nodes, providing a *Logic Density Diagram* (LDD). ADD can provide an optimal perspective of the system relatively to aspect directly or indirectly related to traffic, communication, large scale, routing; LDD results more direct if related to the evaluation of tasks, gateway service and hierarchical organization. As general assumption, the global point of view provided by both these diagrams could result as important in the phase of evaluation/optimization that should result as simplified. The minimization of MD-eval (resource optimization) can or cannot have a direct relationship with Power; the proportional decreasing of Power represent probably the desired situation; however, in a great number of situations, resource optimization can result in a better organization and management of resource that has a little (or not appreciable) effect over Power but a considerable impact on the improvement of system time-life.

5 Simulation

Three different scenarios will be analyzed. These architectures should be considered as examples but, at the same time, they represent some of most common and relevant topologies in WSN deployment context. Figure 4,a (Basic Cell, BC) represents a multi-sink architecture composed by a number of basic cells formed by a sink and several sensor nodes. Communication into the cell is mono-hop and nodes simply provide independent sensor tasks and so its role is always considered as active. Every zone has its own keep-live system to detect eventual failure situations. Sinks are supposed as to be organized in hierarchical way and provided with both high and low power (and range) communication capacity.

Configuration	Communication	Sensing	Actuator	QoS
BC	Mono-Hop	Independent	No	1%
HDC	Multi-Hop	Independent	1-5Hops	5%
CWSN	Multi-Hop	Independent/ Cooperative	1-2Hops	6%

Table 1. Simulation Configuration.

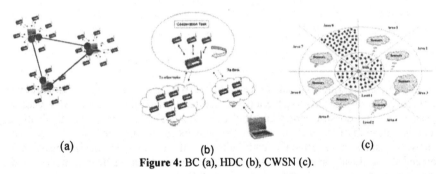

| (a) | (b) | (c) |

Figure 4: BC (a), HDC (b), CWSN (c).

The second scenario, showed in Figure 4(b), is the High Density Cell (HDC). HDC is characterized by a unique sink and by a great number of sensor nodes uniformed distributed into considered area. Communication is multi-hop and each sensor node has really two important roles: it is part of communication network as transit node (passive role) and it has an independent sensor task (active role). Considering topology, a more advanced Keep-live system, and consequently characterized by a higher resource request, is provided too. The last scenario (Figure 4,c) shows Cooperative WSN (CWSN) that proposes a mono-sink architecture with the same characteristics of HDC but provided with Sensor Task Structures [5]; these mechanisms provides the system with co-operative sensing that guarantees great accuracy in measurements and an important support for fault detection and related reconfiguration. In CWSN architecture, only a limited number of sensor nodes are part of cooperative sensor tasks; the others provide an independent sensing service. Results are showed in Figure 6 (Power), Figure 7 (contextualized analysis) and Figure 9 (MD-eval). If analysis is limited to classic approach (energy consumption evaluation that is represented by Power in proposed model), HDC and CWSN result `as no distinguished and, moreover, the knowledge related to network activities results as very limited.

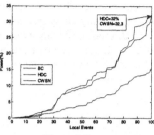

Figure 6: Power.

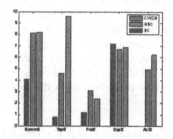

Figure 7: Network Influence/Node role

An extended analysis, that considers power consumption both with a model for the evaluation of network influence (Figure 7) on local resource and, consequently, the determination of the node "role" in the network, could represent a key issue to resource evaluation and optimization of relationship between performance and resource request. Measurements related to BC architecture analysis detect limited power expense (Figure 6) and optimized communication (Figure 7). These issues are the main motivations that advise hierarchical organizations (Figure 4,a). If this architecture is compared with another that guarantees the same service in the same area, gateway nodes (base stations in this case) properties must be considered too.

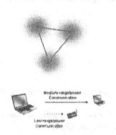

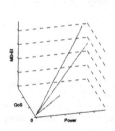

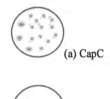

(a) CapC

(b) ProC

Figure 10: BC analysis.
CommC density diagram.

Figure 9: MD-eval.

Figure 8: CWSN analysis.

The best and simple way to provide information in this sense is to consider CommC density diagram (Figure 10). In the few cases in which base station is oriented to limited resource too, this configuration could result as inappropriate. The diagram shows the real "architecture" cost: sensor gateway nodes have the real ownership of communication. This last point must be considered at the time of a comparison. Figure 7 shows a great passive role for sensor nodes of HDC architecture. Geographic nodes distribution over considered area is uniform (Figure 4,c) and so nodes next to sink are really very loaded (Figure 5). TopC density diagram (Figure 11) shows this role considered as too much passive. Moreover, energy consumption is directly related with communication and so the power of nodes next to sink rapidly decreases and could rapidly isolate certain parts of network with very negative impact over network lifetime. This advises a topology with no uniform geographic distribution: nodes density should be inversely proportional with sink distance and routing protocol strategy should be oriented to scalability. A combined analysis of CapC (Figure 8,a) and ProC (Figure 8,b) density diagrams related to CWSN permits to have a map of activities (sensing in this case). If analysis is limited to sensing activities (CapC), cooperative sensor tasks are no localizable or localizable with difficulty. As showed, integrating the analysis with information related to data process (activity of sensor), the localization results very easy.

6 Conclusions and Future Work

An approach for analysis of modern WSNs resource in function of architecture, desired characteristics and related constraints was proposed. Theoretical assumptions and models were related with characteristics of concrete sensor nodes and with realistic and consistent scenarios. Analysis technique proposed results as effective considering both elementary and complex situations; moreover, it can optimally support the designer in presence of critical tradeoffs and challenges. Simulation results can be considered as in accordance with theoretic assumptions. Future extensions of model are mainly related to the explicit consideration of large scale contexts and, mainly, to the integration of proposed models with a well-defined QoS model that should explicitly represent the relationships between QoS mechanisms and requirements on both local and global resource.

References

1.   I. Akyildiz, W.Su, Y.Sankarasubramaniam, E.Cayirci, *A survey on Sensor Network*, IEEE Communication Magazine, August 2002.
2.   Dazhi Chen and Pramod K. Varshney, *QoS Support in Wireless Sensor Networks: A Survey*, International Conference on Wireless Sensor Network 2004, Las Vegas, Nevada, USA.
3.   CrossBow Technology, www.xbow.com.
4.   ZigBee Alliance, www.zigbee.org.
5.   M. Ilyas, and I. Mahgoub, *Handbook of Sensor Networks: Compact Wireless and Wired Sensing Systems*, CRC Press, 2004.
6.   Yuanli Wang, Xianghui Liu, Jianping Yin, *Requirements of Quality of Service in Wireless Sensor Network*, International Conference on Networking, International Conference on Systems and International Conference on Mobile Communications and Learning Technologies (ICNICONSMCL'06).
7.   Dazhi Chen and Pramod K. Varshney, *QoS Support in Wireless Sensor Networks: A Survey*, International Conference on Wireless Sensor Network 2004, Las Vegas, Nevada, USA.
8.   Jan Beutel, Matthias Dyer, Lennart Meier, Matthias Ringwald, Lothar Thiele, *Next-Generation Deployment Support for Sensor Networks*, 2nd International Conference on Embedded Networked Sensor Systems, Baltimore, MD, USA, November 2004.
9.   Vassileios Tsetsos, George Alyfantis, Tilemahos Hasiotis, Odysseas Sekkas, and Stathes Hadjiefthymiades, *Commercial Wireless Sensor Networks: Technical and Business Issues*, Second Annual Conference on Wireless On-demand Network Systems and Services (WONS'05).
10. S. Tilak, N. B. Abu-Ghazaleh and W. Heinzelman, *Infrastructure Tradeoffs for Sensor Networks*, WSNA02, September 2002, Atlanta, Georgia, USA.

A WSAN Solution for Irrigation Control from a Model Driven Perspective

Fernando Losilla, Pedro Sánchez, Cristina Vicente-Chicote, Bárbara Álvarez, Andrés Iborra

fernando.losilla@upct.es
Universidad Politécnica de Cartagena (Spain)

Abstract. Wireless Sensor and Actor Networks (WSAN) constitute a growing research field in several engineering areas. One very interesting domain of WSAN application is precision agriculture, and in particular, the automatic control of tree irrigation depending on sap flow levels. Nowadays, the software development process followed in these kinds of applications is largely dependent on the platform where the final implementation is done. Consequently, commonly desired attributes such as flexibility, reuse and evolution are relegated to a second level of priority. Nevertheless, the growing interest in WSAN has been led to advances from different points of view: new application domains, new middleware, new simulation environments, and so on. In spite of all these advances, WSAN development is today needed of concrete mechanisms that make easy the software generation process. This paper summarizes our contribution in this field from two points of view: as an agronomic solution and as new opportunities for affording the construction of these systems taking into consideration the most recent advances in software engineering.[1].

Keywords: Model driven approach, Domain Specific Languages, Precision Agriculture.

1. Introduction

Recent technological advances have led to the emergence of wireless sensor and actor networks (WSAN). These networks are able to observe the physical world, process data, make decisions on the basis of these observations, and carry out concrete operations on the environment.

WSAN applications constitute a new way of acquiring data and controlling the physical world, and therefore they are very useful for developing a broad range of

[1] This work was partially supported by the Spanish CICYT project MEDWSA, reference TIC2006-15175-C05-02 and the Regional Government of Murcia Seneca Program, reference 02998-PI-05.

Please use the following format when citing this chapter:

Losilla, F., Sánchez, P., Vicente-Chicote, C., Álvarez, B., Iborra, A., 2007, in IFIP International Federation for Information Processing, Volume 248, Wireless Sensor and Actor Networks, eds. L. Orozco-Barbosa, Olivares, T., Casado, R., Bermudez, A., (Boston: Springer), pp. 35-46.

applications such as [1]: environmental monitoring, health control by tele-medicine, precision agriculture, military operations, transport tracking, etc.

Besides, model driven software development (also known as model driven engineering or MDE [2][3]) solves the common problem of the strong dependence of the software process on the final execution platforms. The MDE approach focuses on the use of models through which software applications can be described while taking independent platform concepts into account and promoting the definition of semi-automated transformation mechanisms from those models in final code.

These approaches to the development of WSAN based applications entail the following research activities: (1) to adopt and apply a method that guides the domain engineering process from which the concrete product line study will be derived; (2) to define a domain-specific language with which models can be produced by representation of concepts from that domain; (3) to select an execution infrastructure-independent language with which all architectural design decisions for the product family can be represented; (4) to define the transformations between models taking into account the meta-models (that is, models that define a language for expressing other models) identified for each language utilized in the process; and (5) to develop tools which provides support for the whole process.

In this paper we describe with detail the adopted solution for a well know problem of precision agriculture: control irrigation by measuring the tree sap flow. This example has allowed us to put in practice the languages, tools and developed software artefacts.

For a good understanding of the work, we first describe the case study in section 2. Section 3 gives an introduction to the concepts that could enhance the development of WSAN applications: the model driven software development perspective. Sections 4 and 5 present the case study solution from a model management point of view. Section 6 gives the related work and section 7 conclusions and future works.

2.   Case study description

The MITRA test bed consists of thirty TinyOS-based nodes (see Figure 1) deployed in an almond orchard located in the region of Murcia (southeast of Spain), where the climate is semiarid Mediterranean. Given the shortage of water in this region, the prime objective of the system is to regulate tree irrigation according to water stress, that is, to water the trees only when required. Water stress is measured using the heat pulse compensation method [4] which consists in generating a pulse of heat throughout an axial line of the tree trunk using a resistor element. Then the temperature evolution is measured above and below this line to determine the sap flow and hence the water stress.

Figure 2 shows the hardware architecture of these MITRA nodes. The microcontroller, together with a driver, generates an electrical pulse and applies it on the heater. The heater is a very thin resistor made up of a nicron wire inside a 2 mm steel tube. To calculate the sap flow, it is necessary to measure the temperature at different depths of the tree trunk. To get this information, MITRA nodes provide two temperature sensors. In addition, MITRA nodes are equipped with a port for connecting up to eight additional sensors (e.g. luminosity, temperature, humidity of the soil, and air pressure).

The microcontroller processes the signals provided by the temperature sensors in order to calculate the sap flow. This datum is reported via WiFi every three hours (for energy-efficiency considerations) to a simple PC, which controls the irrigation process on the basis of the information received from all the deployed MITRA nodes.

When the PC detects that some water is needed, it sends a "start irrigation" order to the watering system via WiFi, and conversely, when it detects that the trees are sufficiently watered, it sends a "stop irrigation" order. The section 4 explains how the MITRA system was implemented using the tools and the methodology presented in this paper.

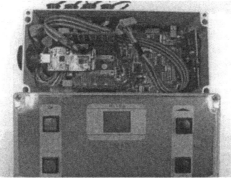

Figure 1: The MITRA node.

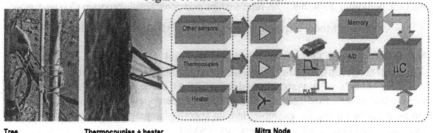

Figure 2: MITRA node hardware architecture.

3.   A model driven perspective for developing WSAN applications

Model Driven Engineering (MDE) promotes software development centred on the concept of a systematic use of a *model*. In this approach, models play the most important role because they are the chief artefact guiding not only the development and documentation of the software but also its management and evolution. In MDE [2] models are created on the basis of formal *meta-models* which describe complementary views of a system at different levels of abstraction.

In MDE, software can be described by means of models at high levels of abstraction, and then these models can be compiled into other representations closer to the implementation level (the executable code) by applying predefined transformations. Because formal descriptions of the meta-models are used, both the building of models and the transformations between them can be done using tools that raise the level of software process automation.

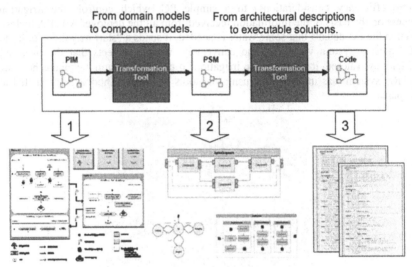

Figure 3: Model Driven Engineering using MDA abstraction levels.

The MDA (Model Driven Architecture) proposal [5] from the OMG (Object Management Group) fits the MDE perspective. MDA proposes three abstraction levels for models of a system (see Figure 3). At the highest level, the platform-independent model (PIM) gives a platform-independent view of the system. This level can be described using for example domain specific languages. The successive lower levels capture platform-dependent features (PSMs).

These PSMs may or may not correspond directly to the code level. MDA promotes the description of software by means of PIM models unrelated to any concrete technology. In a subsequent step, the user selects the final implementation platform to which the upper models will be converted. MDA proposes a combination of the above approach with standards such as MOF (MetaObject Facility), UML (Unified Modelling Language) and XML (Extensible Markup Language) in order to achieve interoperability among the tools involved in the software development process. The MOF specification defines a language and a set of standard interfaces that can be used to define and manipulate a set of meta-models and their corresponding models.

There are several commercial tools for creating, accessing and modifying these meta-models through MOF interfaces. The most representative is the Eclipse Modelling Framework, which has been the selected tool for supporting our work.

A Domain Specific Language (DSL) [6] is a language that offers very good expressive power to capture the requirements of a particular domain at a PIM level. There are many advantages to considering languages of this kind. A DSL provides concrete abstractions to represent concepts from the application domain and offers a natural syntax notation and mechanisms for optimization and error checking.

One of the key factors when defining a DSL is its ability to provide users with a set of modelling primitives very close to the domain abstractions. The DSL we have defined provides concrete and precise constructs for defining WSAN applications on an independent implementation platform such as MDE promotes.

This WSAN-DSL by the authors (an example in next section) provides constructs for specifying both structure and dynamics of WSAN applications at two different levels: node level and region level. It considers a WSAN as an aggregation of regions where a region is seen as a set of atomic nodes and sub-regions with similar characteristics.

The use of regions allows for the building of heterogeneous WSANs where not all nodes have to implement the same functional subsystems (i.e. localization, routing and other specialized processing), but they can be grouped into regions according to what subsystems they implement and how they do it.

Figure 3 summarizes the approach adopted. The picture depicts the separation between PIM and PSM at different abstraction levels. Models represented at a certain level are then transformed into models at lower abstraction levels by means of tools which supply the process with automatic support.

Each level contains a representation of the system using a meta-model that provides formal support for a well-defined modelling or implementation language (model management perspective). Next section presents the obtained models for the case study taking into account the referred software artefacts.

4. General description of the solution

The first step in the proposed methodology for WSAN application development is to build a model of the target system using the WSAN DSL. This high level of abstraction modelling language (meta-model) provides all the concepts and relationships commonly used for specifying WSAN applications.

As noted earlier, a new graphical modelling editor which allows WSAN domain experts to graphically describe the structure and the behaviour of their systems has been developed on the basis of this WSAN meta-model. The MITRA system model depicted using this DSL is shown in Figure 4. Two different regions have been defined in this model. The first region includes two node groups, one representing the MITRA nodes deployed in the almond orchard (SAP Monitoring NodeGroup) and another representing the irrigation control node (NodeGroup containing only one node). The second region contains only one NodeGroup with a single node (Sink node) with the responsibility of transferring data between the WSAN and a computer.

The behaviour of the three different types of nodes (node groups) defined in the model is specified following a three-step process: (1) select the sensors to be read, (2) select the node activities (from those provided by the WSAN DSL), and (3) link all these elements together according to the rules specified in the meta-model.

As can readily be appreciated from the figure, the behaviour of the SAP Monitoring NodeGroup contains two entirely uncoupled activity sets, one describing the sensing loop and another describing how the collected data are sent to the Sink NodeGroup via WiFi. The model transformation which translates WSAN DSL models into generic component models uses this information to place uncoupled activity sets (detected in the WSAN DSL model) into orthogonal state regions in the transformed generic component model (Figure 3, step 2).

Finally, the messages required to model component communication are also inferred. For instance, the MITRA system requires two different messages, one containing the sap values sent from the SAP Monitoring NodeGroup to the sink node and another containing the irrigation orders the PC sends to the Irrigation Control NodeGroup.

Once the generic component model has been obtained from the initial specification, a new model-to-model transformation is performed to obtain an equivalent NesC component model. This transformation identifies certain component patterns in the original model and creates the corresponding NesC components, configurations (complex components), and interfaces. The resulting Nesc model is then automatically transformed into NesC code, using the MOFScript model-to-text transformation implemented to support the last step of the proposed methodology.

The final application that emerges at the end of this process has been tested under real conditions (in an almond orchard belonging to the School of Agricultural Engineering of the Technical University of Cartagena, Spain), with successful results. Obviously, the final application code is still far from being optimized since the tools, particularly those relating to model transformations, still need improving. However, this case study has proven the viability of the approach, demonstrating a significant reduction in the effort required for development.

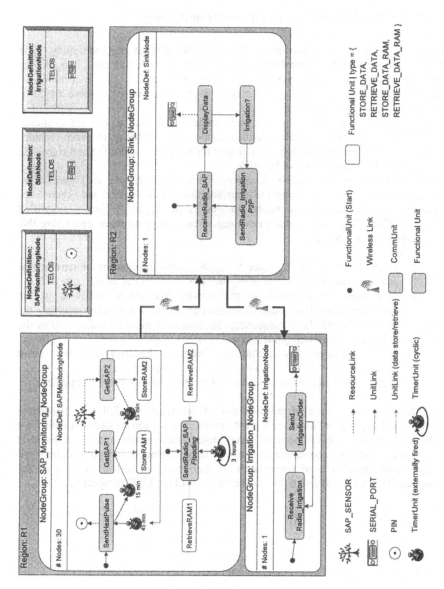

Figure 4: MITRA System model depicted using the WSAN DSL graphical modelling editor.

5. More detail about the transformation process

Figure 4 shows a DSL-description of the functionality of the case study. This representation needs to be translated into a more detailed one that takes into account the services provided and required by each device included in the description. An analogous description of part of the system is given in Figure 5, where an UML Activity Diagram has been used.

Following the UML notation, sent and waited events, and clocks are explicitly represented. It has been considered a unique super-state for the motes which includes two concurrent activity descriptions (left side for sap flow monitoring and right side for periodical sending of data with radio).

It is possible to deduce the component structure described in Figure 6 by considering the set of components and their interfaces. Both Figure 5 and Figure 6 are the input for the compilation process to NesC code. Figure 7 gives a simplified instance of the final code. Some rules need to be considered when generating this code could be the followings:

1. Each referred clock in Figure 5 implies a basic mechanism of starting (for instance, code #29) and attending the associated timeout event (code #53).
2. The sequence described in the activity diagram is followed in the implementation. For example, Sensor.GetSAP1() is done after receiving the end of Timer2. In consequence, the command must be called within the handling of the event (code #43).
3. It is necessary to distinguish different handlers for the event Sensor_Receive_Data with a local variable (code #49...#51).
4. Commands init() and start() must include the initialization of components.

Although we are aware of the difficulty of generating the full code needed to compile the application, however a great percent of code can be deduced both from activity diagrams and components descriptions. Good results can be obtained if appropriate rules are considered in the tool implementation and there exists well documented repositories of components. In this sense, we follow working on the mapping between the different levels of abstraction of the MDE perspective.

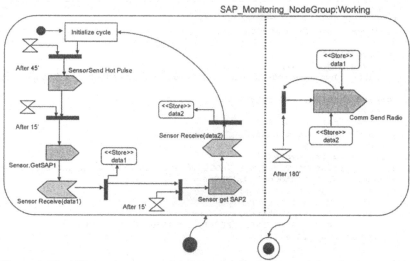

Figure 5: The MITRA System model represented as UML Activity Diagrams inside the state 'working' of the node SAP_Monitoring_NodeGroup.

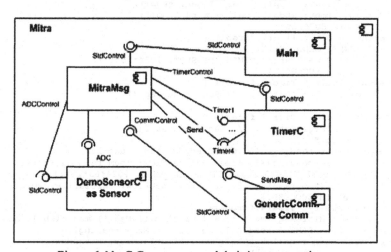

Figure 6. NesC Components and their interconnections.

```
1. includes MitraMsg;
2. module MitraM
3. {uses {
4.  interface StdControl as CommControl;
5.  interface SendMsg as Send;
6.  interface Timer as Timer1; // First 45 minutes clock
7.  interface Timer as Timer2; // First 15 minutes clock
8.  interface Timer as Timer3; // Second 15 minutes clock3
9.  interface Timer as Timer4; // three hours clock
10. interface StdControl as TimerControl;
```

```
11. interface ADC;
12. interface StdControl as ADCControl;}
13. provides {interface StdControl;}
14. }
15. implementation
16. {TOS_Msg msg_global;
17.  uint8_t num_reading=1; uint16_t reading1, reading2; // sap flow
values
18. command result_t StdControl.init() {
19.    call ADCControl.init();
20.    call TimerControl.init();
21.    call CommControl.init();
22.    return SUCCESS; }
23. void initialize_cycle(){
24. call Timer1.start(TIMER_ONE_SHOT, 1024*60*45); // 45 minutes}
25. command result_t StdControl.start() {
26.    call ADCControl.start();
27.    call TimerControl.start();
28.    call CommControl.start();
29.    call Timer4.start(TIMER_REPEAT, 1024*60*60*3); //3 hours
30.    initialize_cycle();
31.    return SUCCESS; }
32.
33. command result_t StdControl.stop() {
34.    call ADCControl.stop(); call CommControl.stop();
35.    call Timer1.stop();  call Timer2.stop();
36.    call Timer3.stop(); call Timer4.stop();
37.    return SUCCESS;}
38. event result_t Timer1.fired() {
39.    SENSOR_SEND_PULSE();
40.    call Timer2.start(TIMER_ONE_SHOT, 1024*60*15);
41.    return SUCCESS;}
42. event result_t Timer2.fired() {
43.    call ADC.getData();  // Sensor.Getsap1()
44.    return SUCCESS;}
45. event result_t Timer3.fired() {
46.    call ADC.getData();  // Sensor.Getsap2()
47.    return SUCCESS;}
48. async event result_t ADC.dataReady(uint16_t sapFlow) {
49.    if(num_reading==1) {reading1=sapFlow; (num_reading++%2)+1;
50.        call Timer3.start(TIMER_ONE_SHOT, 1024*60*15);}
51.    elsif(num_reading==2) {reading2=sapFlow; initialize_cycle();}
52.    return SUCCESS;}
53. event result_t Timer4.fired(){
54.    struct MitraMsg *message = (struct MitraMsg *) msg_global.data;
55.    message->RAM1 = reading1; message->RAM2 = reading2;
56.    call Send.send(TOS_BCAST_ADDR, sizeof(struct MitraMsg),
57.        &msg_global));
58.    return SUCCESS;}
59. event result_t Send.sendDone(TOS_MsgPtr msg, result_t success){}
60.}
```

Figure 7. Code for the MITRA Application

6.   Related work

The foregoing sections showed the benefits of applying a model-driven approach to WSAN application development. The results have been demonstrated by considering a precision agriculture case study. Similar experiments have demonstrated its viability for other reactive systems. But few work-patterns related to this topic have been found in the WSAN domain and none of them use MOF as the underlying substrate. GRATIS II [7], built with the Generic Modelling Environment (GME) meta-modelling tool, constitutes a very well known approximation to the problem, but it does not cover the whole development process. It uses a NesC meta-model and can be used to generate TinyOS configuration components from graphical models and *vice versa*. ATaG [8], also developed with GME, offers a data driven approach for end-to-end application development. In order to model an application, a set of abstract tasks representing types of information processing functions in the system is used along with a set of abstract data items that represent types of information exchanged between abstract tasks. The programmer must supply the code for the implementation of each abstract task and each abstract item of data, but no mechanism is supplied to model this behaviour. Another environment which uses meta-modelling techniques is CADENA [9]. With CADENA, software can be generated not only for WSAN but for any type of application with a component-based architecture, but this generality makes the WSAN software development process excessively complex since domain particular concepts are not used in the functionality description.

The work presented in this paper addresses the creation of a new environment to support the whole WSAN software development process. The resulting environment aims to greatly simplify the construction of applications, making it possible to describe network node behaviour with a small set of concepts relating to the WSAN domain while hiding implementation details from the developer. The application description is currently transformed to TinyOS components in NesC language, but a transformation to other node-level programming environments may be possible provided that they offer a textual or MOF-compliant language for application specification and that they offer an event-driven or thread-based programming model, scheduling schemes, fault tolerance, etc.

7.   Summary and future work

This paper has introduced a new model driven method for the development of WSAN applications that allows for the use of a DSL, meta-models and transformations between models. The description of our results has been supported by a WSAN case study. On the one hand MDE offers automation of the development process and platform independence, while on the other hand the study of the WSAN domain approach offers the best possible reuse in software development. Families of software products are organized around an architecture that utilizes the common features of their members and hence offers a set of reusable artefacts. Moreover, a model-driven software development process focuses on defining models independently from platforms; it also focuses to a greater or lesser extent on the automation of model compilation, and finally on code generation.

Most domain engineering proposals place a great deal of stress on modelling activities as a means of developing product families. The key idea of domain specific languages is to be able to represent the concepts of the application domain directly. In this sense domain specific languages have several advantages over other kinds of general-purpose languages, basically because [6] the constructs are closely related to the concepts of the domain, and these languages usually provide a graphic notation and specialized tools (editors, optimizers, etc.) which gather a great deal of information from the domain. They also provide the opportunity to automate much of the process.

This makes for considerable improvement in the development of reactive systems, particularly in the domain of WSAN applications. We follow working on the implementation of the mentioned tools, the clear definition of transformation rules between models, the demonstration of MDE benefits (such as the independence from the final execution infrastructure) and, finally, the definition of metrics to know the WSAN application time reduction.

References

1. K. Römer, F. Mattern, "The design space of wireless sensor networks", IEEE Wireless Communications, pp. 54-61, Dec 2004.
2. S. Kent, "Model Driven Engineering", in Proc. of Integrated Formal Methods: Third International Conference, IFM 2002, Lecture Notes in Computer Science, vol. 2335, Springer-Verlag, 2002.
3. S. Deelstra, et al., "Model Driven Architecture as Approach to Manage Variability in Software Product Families", in Proc. of the Workshop on Model Driven Architecture: Foundations and Applications (MDAFA 2003), pp. 109-114, University of Twente, June 2003.
4 P. Becker, "Limitations of a compensation heat pulse velocity system at low sap flow: implications for measurements at night and in shaded trees", Tree-physiol. Victoria [B.C.] Canada. Heron Pub., Mar 1998 v.18(3) p.177-184.
5. OMG Model Driven Architecture (MDA) Guide v1.0. Available at: http://www.omg.org/docs/omg/03-06-01.pdf
6 J. Estublier, G. Vega, A. D. Ionita, "Composing Domain-Specific Languages for Wide-Scope Software Engineering Applications", in Proc. of MoDELS 2005, LNCS 3713, Springer-Verlag, 2005, pp. 69-83.
7. P. Volgyesi, et al., "Software Composition and Verification for Sensor Networks", Science of Computer Programming (Elsevier), 56 (1-2), pp. 191-210, April 2005
8. A. Bakshi, et al., "The Abstract Task Graph: A Methodology for Architecture-Independent Programming of Networked Sensor Systems", in Proc. Workshop on End-to-End, Sense-and-Respond Systems, Applications, and Services (EESR'05), Washington, USA, June 2005.
9. G. Trombetti, et al. "An Integrated Model-driven Development Environment for Composing and Validating Distributed Real-time and Embedded Systems", in Model-Driven Software Development, S Beydeda, M. Book and V. Gruhn (Eds), Springer-Verlag, 2005

An Action Activated and Self Powered Wireless Forest Fire Detector

Johan Sidén[1,2], Andrey Koptyug[2], Mikael Gulliksson[2], and Hans-Erik Nilsson[1]

[1] Mid-Sweden University, Electronics Design Division,
851 70 Sundsvall, Sweden
{Johan.Siden, Hans-Erik.Nilsson}@miun.se

2 Sensible Solutions Sweden AB, Storg. 90
851 70 Sundsvall, Sweden
{Andrey.Koptyug, Mikael.Gulliksson}@sensiblesolutions.se

Abstract. Placing fire detectors in a forest is usually associated with powering problems since the sensors do not have access to external power supply and a periodical change of internal batteries is an undesired option. This paper presents an approach where the fire sensor itself, when heated by nearby fire, generates electrical energy to power a radio transmitter. Presented energizing fire sensor is environmentally friendly and can be mass produced at a very low cost. Upon activation the sensor produces enough power necessary to operate most standard radio transmitters depending on what communication system is chosen for operation. The fire detector unit can be deployed from either helicopter or manually from the ground. The sensor can be designed to activate itself at different temperatures to suit different climate zones. Rough guidelines are given for estimation of attenuation of radio wave propagation in forest areas in order to predict maximum transmit distance.

Keywords: Forest Fire Detectors, Action Activated Tags, Energizing Sensors

1 Introduction

Forest fires, particularly those occurring in dry windy weather pose serious risks. Such fires can cause significant deforestation, resulting in the loss of wild life and its habitat, seriously changing ground water levels and river flows, changing microclimate etc. Private and industrial properties are also damaged and it can result in the loss of human lives. As populations and the attractiveness of "green living" increase, forests are being subjected to more and more exposure to human activities, which in turn, can lead to the triggering of forest fires.

In order to avoid the serious consequences associated with forest fires, improvements in fire prevention, detection, monitoring and fighting are constantly being carried out. It is commonly accepted that forest fire prevention and its detection at the earliest possible stages are the most critical exercises. Significant progress is currently being achieved in all the above mentioned areas, but fire detection and progress monitoring are more dependent on technology than are the preventive measures.

Please use the following format when citing this chapter:

Sidén, J., Koptyug, A., Gulliksson, M., Nilsson, H.-E., 2007, in IFIP International Federation for Information Processing, Volume 248, Wireless Sensor and Actor Networks, eds. L. Orozco-Barbosa, Olivares, T., Casado, R., Bermudez, A., (Boston: Springer), pp. 47-58.

Early forest fire detection and precise localization significantly reduce both the problems and costs associated with fire fighting because of the increased efficiency involved in tackling the fire. At the same time, there should be a corresponding system able to provide real time information with regards to the initiation and progress of a fire which is neither too complex nor excessively expensive [1]. Several approaches to achieve this have been suggested, some of which are already in use at the present time and involving the use of remote sensing, area imaging and laser-based detection. This paper proposes a different approach, based upon wireless sensor-nodes with energizing sensor elements to power the radio unit.

2 Existing Fire Detection systems

Area imaging systems are either based on stationary ground towers [2], [3], or are air-born [4] or satellite based [5]. In this instance, fire detection is based on the computer analysis of the images, often acquired in multiple spectral bands [6]. In recent years, satellite imaging in forest fire detection has received increasing attention, primarily due to its possibility to scan large areas. Significant progress has been achieved in this case, through the development of better imaging sensors [7] and also in improvements to image processing [8]. At the same time, complex processing algorithms may require both very high resolution images and significant processing time, which makes real-time fire detection an impossibility. Airborne (airplane and helicopter based) imaging systems are commonly less expensive than satellite ones, but they require far more complex management and cannot compete with satellite imaging in either area coverage or the continuity of monitoring.

Imaging systems also often lack the necessary resolution and sensitivity in cases involving thick clouds and thick smoke generated by the fire itself. Laser-based systems, referred to as LIDAR type [9] help to overcome these problems. However, laser beams are very narrow and almost non-diverging and in order to provide adequate area coverage using a single LIDAR unit, it should be operated in a sweeping mode, as is the case with RADAR. Such mechanical systems, capable of fast angular LIDAR area sweeping, can be complex and rather expensive. Another serious disadvantage of the above optical and laser based systems is that they not only require constant technical attention and thus could not be easily maintained in inaccessible areas, but also utilize a serious amount of power in order to operate.

Small stationary sensors provide an attractive alternative or a valuable addition to the systems involved in the early detection of forest fires. Such sensor systems can provide almost instantaneous fire detection combined with very high localization precision. Wired fire sensor networks commonly used inside buildings are not really applicable in this case because of the costs and technical difficulties involved in covering large forest areas. Consequently, to be effective such networks must use radio communication. Thus, each sensor should not only be able to detect the fire but also be able to send this information to the nearest monitoring station.

Because such sensors must be permanently deployed in the forest, certain preconditions must be fulfilled. Ideally such system should meet following requirements:

- The system should be able to cover large forest areas;
- The system should be able to recognize which sensor has produced the alarm signal (localization);
- Sensor units should be relatively small and easy to deploy (includes deployment from air);
- The sensor units should have a very long lifetime and, once deployed, should be maintenance free for several years;
- Sensor units should not contain materials or components potentially dangerous to the environment, wild life or humans;
- The sensors should not emit dangerous chemicals when destroyed by fire.

Presently, the most promising area is considered to be that utilizing the concept of networked sensor nodes, which either use existing wireless networks such as GSM/GPRS or through the formation of separate dedicated networks, including those based upon TCP/IP [10]. In the case with TCP/IP, the individual sensor node is not forced to radio-communicate its signal directly to a monitoring station, but only to the nearest re-broadcasting node. This assists in significantly reducing the transmitter output power and thus enabling it to be kept within specified limits.

Mentioned existing wireless sensor node solutions are rather bulky and use heavy-duty batteries which somewhat limits their application in forest fire detection systems since such units require regular service such as battery replacement and are relatively expensive. They must also be deployed in accessible places and contains significant amounts of potentially harmful chemicals (the battery, solder, plastics etc.). Promising concepts involving power harvesting from various environmental processes like light, vibrations, etc. [13] has not yet yielded a satisfactory solution which enables maintenance-free power sources for such sensor nodes, but has been able to significantly prolong the lifetime of traditional batteries.

This paper presents a solution, based on the new maintenance-free Action- Activated Tag (AAT) concept. Upon activation, this sensor provides sufficient electric power for a radio-transmitter to operate before being completely destroyed by the triggering fire.

3   The Action Activated Tag Concept for Forest Fire Detection

The Action Activated Tag concept refers to systems using the detected physical quantity (in this case heat) to produce or release sufficient power to operate the necessary electronics. These sensors can be compared with printed dry batteries and were originally developed for wetness detection in incontinence diapers where a sufficient amount of wetness powered a low-power radio transmitter when a diaper required changing [16]. These sensors were modified for use as secondary sensors in a fire detection system. In this case, raising temperatures triggers the release of a small amount of water, stored in a small container inside the sensor body. The water activates the small energy cell, providing power for the sensor node electronics. The construction of the discussed fire sensor unit is outlined in Fig. 1.

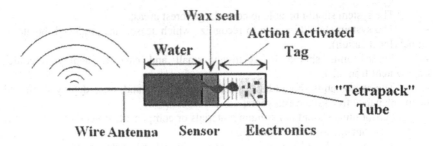

Fig. 1. Functional drawing of the AAT fire sensor node illustrating how water is entering a moisture sensor when a protective wax seal melts under significant heat.

The small water container in the sensor is separated from the sensor element by a plug that melts at a given elevated temperature, allowing water to flow over the sensor, which in turn produces sufficient electrical power to operate the radio transmitter. The construction of the sensor involves printing using special inks and layers of absorbent paper and fibrous materials and is cheap and suitable for bulk manufacturing. The transmit antenna is made of thin wire and the components of the sensor node are encapsulated into an impregnated carton tube, using technology similar to that the food industry uses for juice and milk packages.

The AAT sensor node is constructed in such a way as to minimize any possible environmental impact. Paper-based and other fibrous materials are sufficiently robust in order to withstand seasonal weather changes for several years, but will biodegrade in the longer term. The main triggering chemical is ordinary water and the overall metal and plastic content is almost negligible as it is measured in milligrams per sensor. Instead of using a conventional printed circuit board (PCB), the few necessary electrical components are mounted on a bio-degradable wooden chip substrate using silver ink for conductive traces. Thus, if the sensor is burnt in the fire or degrades over time, dangerous substances will not be produced as a by-product.

Fig. 2. Example of a single sensor element consisting of layers of "newspaper grade" paper with zinc- based (top layer) and carbon based ink (bottom layer). Absorbing layer (doped with electrolyte and dried) is placed between the active layers.

If the sensor unit would be tampered with, the thin wire antenna and the sensor casing will both readily break. The wax plug at normal ambient temperatures is solid and mechanically stronger than the water container, why the water does not enter the sensor area if it for example is stepped upon, but will exit the casing in the opposite direction and not result in a false alarm. It is also a necessity that the wire antenna break easily in order to prevent any harm coming to animals accidentally stepping on a fallen fire sensor and an additional aid to prevent false alarms from vandalism.

The sensor cell is assembled by stacking a few individual paper strips together, each printed with inks with different electrical properties. One of the possible constructions of the printed sensor is shown in Fig. 2. Ink is applied to a paper base using traditional rotary screen or flexographic printing which is a common roll to roll manufacturing process. The inks used are mainly mixtures of carbon, zinc and manganese dioxide. The absorbent layers are doped using a non-harmful electrolyte such as citric acid or cooking salt. The sensors have a very long lifetime and start to produce electric power as soon as a few milliliters of water reach the active area [16]. The sensor element illustrated in Fig. 2 has an active area of a few cm^2 and gives about 1.2 V when wet. To increase the delivered power, sensors are made of several cells rolled together into a compact cylindrical shape.

The wax plug material can be selected to melt at pre-determined temperatures. Tests have been carried out using several different wax substances known to be environmentally friendly and melting points from about 50 to just over 100°C have been observed. Other alternatives include natural asphalt and tar materials depending on the required temperature. It is possible, that minor sensor unit modifications would be required for different climatic zones because of significant differences in the highest summer and lowest winter temperatures. Both the sensor and the node electronics are capable of surviving at temperatures well in excess of 100 °C thus providing proper functionality after the wax seal is broken.

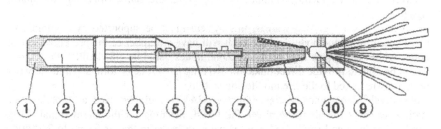

Fig. 3. Cut-through view of the sensor node unit, ready for deployment. (1)- water container with water (2) and wax plug seal (3); (4)- dry sensor element . (5)- paper- based tubular casing; (6)- board with electronic components; (8)- antenna wire, rolled onto the conic rear plug (7); (9)- fluffy tail; (10)- tail fixing plug. See text for further comments.

The AAT sensor node in the present design has a similar formation to that of a small lightweight "rocket" with a fluffy tail that is about 2-3 m long and 2-3 cm in diameter. A cut-through view of the model sensor node unit is displayed in Fig. 3. The nose and tail cone plugs (1 and 7 in Fig. 3) are made of wood. The outer shell (5) is rolled from a few paper layers. Water in a small container (2) is sealed using a wax plug (3) and placed near the rolled moisture-activated sensor element (4). A small board containing the electronic components is fixed in the slot of the rear conic plug (7). To protect the electronics from outer water, it is protected by a layer of environmentally friendly lacquer. The antenna wire, rolled onto the conic part of the rear plug, is pulled out by the fluffy paper strip "tail" (9). The tail is initially fixed in place using a felt plug (10).

To improve the sensor activation, some ethanol (boiling point 78°C) can be added to the water. At seriously elevated temperatures the ethanol vapors and increase the pressure behind the wax seal, assisting it to burst, when it becomes soft. This is helpful in order to inject the liquid into the sensor when the unit is not oriented in a direction that allows injection by gravity. It also assists in the prevention of sensor destruction caused by freezing liquid expansion.

Using the node construction and materials described above it is estimated that it is possible to bring the manufacturing cost per unit down to approximately € 1-3 in the case of mass production depending on the total number of units and choice of radio system.

In order to ensure that the overall system cost remains as low as possible, a simple means of sensor node deployment is necessary. Because AAT sensor nodes are relatively lightweight, they could be deployed from the ground with the assistance of a "slingshot" or similar. At the deployment the coiled thin wire antenna is removed by the flow of air in the fluffy tail and is meant to entangle itself in tree branches or in a bush.

When the sensor node is deployed it may be necessary to record its location, particularly if the monitoring station is at a distance. This is possible by including a short-range and low-cost RFID tag [18] within the sensor unit. The RFID tag's unique ID number and current position is automatically read by an RFID reader within the deployment mechanism and recorded into a database together with aid of a GPS receiver. The same identity number is used when the sensor transmits the alarm signal.

In order to cover significant areas, this system must be supported by a number of more powerful "Zone nodes" where each will have a number of the presented sensor nodes within reach (see Fig. 4). These Zone nodes could be GSM/GPRS based, and would be able to provide direct links into existing networks, or form separate network of their own, linked to the fire monitoring station.

Alternatively, Unmanned Aerial Vehicles (UAVs) could be used as mobile relay stations to scout the air space above the forest, listening for the alarm signals [19]. Modern medium and small UAVs are capable of being in the air for up to 24 hours, constantly maintaining radio links with the "base" directly, via satellite or via GSM/GPRS systems.

An advantage of a UAV-based intermediate system is that one UAV, high up in the sky, is able to cover a much greater area than an earthbound zone node because a radio signal emerging from an alarming sensor is attenuated much less when propagating through air than through forest foliage in an uneven landscape.

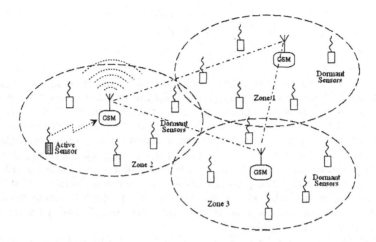

Fig. 4. Simplified hierarchical network structure of the Forest Fire alarm system, based on the AAT sensor nodes.

4 Radio Communication Challenges in Forest Fire Detection Systems

A high sensor unit cost may force the use of a relatively low sensor density in the forest (say, one sensor per several square kilometers), thus reducing the system's capability of detecting the initiation of a fire sufficiently rapidly. If the sensors for example are placed in a square grid with a grid distance of 1 km, and the assumption is made that a fire will spread uniformly in all directions from a starting point, an average circular area of 0.39 square kilometers will be burnt before the sensor unit produces an alarm. A dedicated radio communicating system therefore appears to be more appropriate for the sensor nodes and the use of GSM/GPRS should be used at the zone nodes.

5 Radio Wave Propagation in Forested Areas

There are factors limiting the transmitted power such as the limited capacity of the power source and the legal restrictions for the used communication band, such as those put up in the "free bands" found in [20]. Maximizing the alarm detection distance is therefore of importance. Additionally, in the forest environment the signal will be attenuated by the foliage and branches, further reducing the communication distance.

The most basic formula for the estimation of received power P_r from a unit transmitting with power P_t at a distance R and at a frequency f is the *Friis Transmission Formula* [21]:

$$P_r = \frac{P_i G_i G_r \lambda^2}{(4\pi R)^2} \qquad (1)$$

where λ is the free space wavelength, $\lambda = c/f$ and $c = 3 \cdot 10^8$ m/s is the speed of light. G_t is the gain of the transmitting antenna and G_r the gain of the receiving antenna. Gain is a measure of how an antenna directs its radiated and received power. The preference for this specific application is that an activated alarm unit transmits with equal power in all directions, as it is not possible to predict the spatial orientation of the antenna in a sensor dangling in the trees. The receiving antenna should also have similar properties and should be of circular polarization to minimize the risk of a polarization mismatch. The proposed linear sensor antenna radiates almost omni-directionally, acting either as a traditional half- or quarter-wavelength dipole- or monopole antenna with a toroidal directivity pattern [21] and gain factor of approximately $G \approx 1.6$.

Analysis of (1) provides the general conclusion that with a fixed transmitter power P_t available for radio transmission and fixed threshold received power P_r for alarm signal recognition, the lower transmission frequency (larger wavelength) will provide a longer communication distance R. It should be noted that this conclusion is drawn without taking the frequency dependence of the radio waves attenuated by the forest into account. Unfortunately, this conclusion can also not be fully utilized in present form as it assumes that the antenna size is changing with frequency, which is a significant part of the wavelength. In reality, a sensor unit antenna of length greater than 1-2 meters would be impractical since modules with antennas of such length would be hard to deploy. The correct analysis should take into account the change in the antenna efficiency (gain factor) and the attenuation of the signal by the forest and soil at chosen frequency bands allowable within such systems.

To be efficient, this type of antennas must have dimensions close to one quarter- or half wavelength [21]. Also, the ideal shape for a longer antenna of a sensor deployed in a forest and which will be possibly be dangling in the tree branches has an increased chance of becoming distorted (i.e. not being straight). Corresponding correction factors for the system performance in such a case could be taken from [22] where the performance of bent dipole antennas is characterized. There is also the possibility to redesign antennas in order to achieve higher robustness without seriously degrading their performance [23].

Among the frequency bands, allowed by the EU authorities for license-free transmissions are 13.56 MHz ($\lambda/_4 \approx$ 5.5 m), 27 MHz ($\lambda/_4 \approx$ 2.25 m), 433 MHz ($\lambda/_4 \approx$ 17 cm) and 865-869 MHz ($\lambda/_4 \approx$ 8.7 cm) [20].

Friis Transmission Formula does not take into account additional signal losses caused by the environment of the antenna, however many models exist which allow for this scenario. One of the simplest, but most accurate, models is a *two-ray* model. This model considers two antennas elevated above the ground (transmitter antenna at H_t and receiver antenna H_r), and takes into account one direct radio-signal and one reflected from the ground as illustrated in Fig. 5.

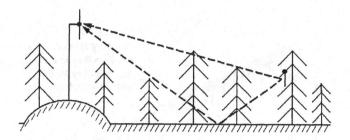

Fig. 5. Two-ray radio propagation model where the activated fire alarm unit is deployed on a tree branch and the alarm unit's antenna is hanging down. The alarm unit transmits a signal that reaches the base station both directly and via reflection in the ground.

In the case where the working frequencies are lower than approximately 50 MHz, the ground wave should be taken into account. In the forest environment the ground wave is actually a "canopy" one, traveling along the tree top line. The propagation of both the direct- and reflected waves is calculated using the Friis transmission formula and the total received power due to the superposition of both signals is expressed in the two-ray model as [24]:

$$P_r = \frac{P_t G_t G_r \lambda^2}{4(\pi R)^2} \sin^2\left(\frac{2\pi H_t H_r}{\lambda R}\right) \tag{2}$$

where H_r and H_t are the receiver and transmitter antenna elevations, respectively, and all other values are as in expression (1).

The second term in expression (2) describes the interference pattern, caused by superimposing two waves with different phase lag. For larger distances $H_t·H_r \ll \lambda·R$ the approximation $\sin x \approx x$ can be used and expression (2) can be simplified as:

$$P_r = \frac{P_t G_t G_r (H_t H_r)^2}{R^4}. \tag{3}$$

Interestingly, but as according to (3), the communication distance R is no longer dependent upon the working frequency (or the wavelength, which is the same) for a fixed transmitter power P_t and the fixed threshold receiver power P_r.

Losses due to the soil and forest foliage can be also taken into account:

$$P_r = \frac{P_t G_t G_r (H_t H_r)^2}{R^4} \cdot e^{\alpha R} \tag{4}$$

In this case it is possible to use the empirically determined attenuating factor $\alpha \approx 0.2$ dB/m, (measurements were carried out for different kinds of forest terrain [25]). Fig. 6 presents a comparison of the propagation attenuation dependence on the distance from the transmitter, as calculated using expressions (1)-(4).

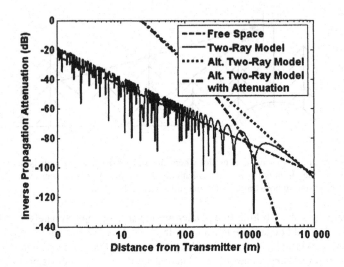

Fig. 6. Attenuation of radio waves transmitted with 2W ERP as modeled using equations (1)-(4). See text for the details.

The attenuation plotted in Fig. 6 was calculated for the frequency f=866 MHz with the transmitter power of P_t=2 W [27]; $\lambda/4$ dipole antennas with G_t= G_r= 1.64; antenna elevations of H_t=10 m, H_r=20 m; attenuation factor α=0.2 dB/m, using equations: (1) {"*Free Space*"}, (2) {"*Two-Ray Model*"}, (3) {"*Alt. Two-Ray Model*"} and (4) {"*Alt. Two-Ray Model with Attenuation*"}.

Modern radio communication integrated circuits for digital communications have the sensitivity of approximately 130 dB or better, thus the communication distance between the sensor and receiving station could be several kilometers.

These very basic and generalizing calculations in fact only consider the received power. A more accurate *link budget* should consider the Signal to Noise ratio (SNR) for the specific modulation scheme, Bit Error Rate (BER), codes for Forward Error Correction (FEC) and, preferable, even more variables that affect the overall system performance. Overviews of more thorough link budget calculations are found in [24] and [29]. Generally, it is however possible to state that as long as an efficient antenna is available, the lower the wavelength, the longer the maximum communication distance. This does not follow from equations (1)-(4) but certain benefits may be gained, and it is characterized for forested areas in for example [25]. A similar rule of thumb holds for the bit rate where the lower this is the more reliable the communication is which is advantageous for this application since a minimum amount of data is required.

Mentioned models offer *estimates* for reliable communication ranges. For a specific application and area, the only means of achieving genuinely reliable numbers is to perform field tests using the designated equipment.

6 Conclusions

A new electronic forest fire detection system built on the concept of distributed sensors, referred to as Action Activated TAGs (AAT), is presented. The system comprises low-cost sensor units, easily deployed from air or ground. The AAT sensors are dormant and start to transmit the alarm signal only when subjected to the significantly elevated temperatures of a forest fire but not when they are tampered with. The presented sensors are able to overcome serious problems which are present when other electronic forest fire detection sensors are used. The sensor units are lightweight, have a significant life span, are environmentally friendly and are inexpensive to mass produce.

Further development of fire detection systems using AAT sensors is required as is additional work on the sensor construction to maximize the communication distance.

References

[1] D. L. Martell, R. S. McAlpine "Forest Fire Detection Economics", *Wildfire Detection Workshop, March 25 - 27, 2003, Hinton Alberta, Canada*

[2] E. Kührt, J. Knollenberg, V. Mertens "An Automatic Early Warning System for Forest Fires", *Annals of Burns and Fire Disasters - vol. XIV - N 3, September 2001*

[3] D. Schroeder "Evaluation of three wildfire smoke detection systems", *FERIC, Vancouver, B.C. Advantage Report, 2004, Volume 5 , No. 24, p.8*

[4] "Firehawk™, Electronic Forest Fire Detection and Management System", *IFFN No. 23 - December 2000, pp. 105-108*

[5] Y. Rauste, E. Herland, H. Frelander, K. Soini, T. Kuoremaki, "Satellite-based forest fire detection for fire control in boreal forests", 1997, *International journal of Remote Sensing V18, N.12, pp. 2641-2656*

[6] Y. Li, A. Vodacek, R.L. Kremens, A.E. Ononye, C. Tang, "A hybrid contextual approach to wildland fire detection using multispectral imagery", 2005, *IEEE Trans. Geosci. Remote Sens. V.43, pp.2115-2126*

[7] J. Leijtens, A. Court, J. Hoegee, "Microbolometer Spectrometer, Instrument and Applications", *Proceedings of The 4th Symposium Small Satellites, Systems and Services, La-Rochelle, France, 20-24 September, 2004*

[8] A. Ononye, A. Vodacek, R. Kremens, "Improved fire temperature estimation using constrained spectral unmixing", *Proc. 10th Biennial USDA Forest Service Remote Sensing Applications Conference, Remote Sensing for Field Users. Salt Lake City, USA, 2005*

[9] A. Utkin, A. Fernandes, A. Lavrov, R. Vilar, "Eye-safe Lidar Measurements for Detection and Investigation of Forest-fire Smoke", *Intl. Journal of Wildland Fire, V13, 2004, pp. 401- 412*

[10] A. Dunkels, T. Voigt, J. Alonso, H. Ritter, J. Schiller, "Connecting Wireless Sensornets with TCP/IP Networks", *Proceedings of the Second International Conference on Wired/Wireless Internet Communications (WWIC2004), Frankfurt (Oder), Germany, February 2004, Springer-Verlag.*

[11] N. Rodriguez, G. Bistue, E. Hernandez, D. Egurrol, "GSM Front-End to forest fire detection", *Proceedings of ISTAS 2000, Roma, September 2000*

[12] R. Kremens, A. Seema, A.Fordham, D. Luisi, B. Nordgren, S. VanGorden, "Networked, Autonomous Field Deployable Fire Sensors", *Proceedings of the 5th International Wildfire Safety Summit, Missoula, MT, USA, November 2001*

[13] M. Philipose, J. R. Smith, B. Jiang, K. Sundara-Rajan, A. Mamishev, S. Roy, "Battery-Free Wireless Identification and Sensing", *IEEE Pervasive Computing, V4, N. 1, Jan 2005*

[14] J. Schiller, A. Liers, H. Ritter, R. Winter, T. Voigt, "ScatterWeb - Low Power Sensor Nodes and Energy Aware Routing", *Proceedings of 38th Hawaii International Conference on System Sciences (HICSS-38 2005), 3-6 January 2005, Big Island, HI, USA*

[15] S. Roundy, P. K. Wright, J. M. Rabaey, "Energy Scavenging for Wireless Sensor Networks : with Special Focus on Vibrations", *Kluwer Academic Publishers, 2004*

[16] H-E. Nilsson, J. Sidén, T. Unander, T. Olsson, P. Jonsson, A. Koptioug, M. Gulliksson "Characterization of moisture sensor based on printed Carbon-Zinc energy cell", IEEE Proceedings of Polytronic 2005

[17] J. Sidén, A. Koptioug, M. Gulliksson, The "Smart" Diaper Moisture Detection System, Microwave Symposium Digest, 2004 IEEE MTT-S International, Vol. 2 pp. 659 – 662, 2004

[18] K. Finkenzeller, "RFID Handbook", ISBN 0-470-84402-7, John Wiley & Sons 2003

[19] Z. Gouqing, L. Chaokui, C. Penggen, "Unmanned aerial vehicle (UAV) real-time video registration for forest fire monitoring", Geoscience and Remote Sensing Symposium, 2005. IGARSS '05. Proceedings. 2005 IEEE International, Volume 3, 25-29 July 2005 Page(s):1803 – 1806

[20] European Telecommunications Standards Institute, ETSI, www.etsi.org, visited 060515

[21] D. K. Cheng, "Field and Wave Electromagnetics", 2nd edition, ISBN 0-201-12819-5, p., Addison Wesley, UK, 1989

[22] J. Sidén. P. Jonsson. T. Olsson, G. Wang, "Performance Degradation of RFID System due to the Distortion in RFID Tag Antenna" , Proceedings of the 2001 11th International Conference "Microwave & Telecommunication Technology" (CriMiCo'2001), IEEE Catalog Number: 01EX487, 2001

[23] J. Sidén, P. Jonsson, A. Koptioug, "Printed RFID tag antennas with minimal performance degradation when bent", in the Proceedings of Asia Pacific Microwave Conference 2003, APMC'03, 2003

[24] L. Ahlin, J. Zander, Digital Radiokommunikation –System och Metoder, *Studentlitteratur*, p. 36, ISBN 91-44-34551-8, 1992

[25] I. Koviics, P. Eggers and K. Olesen, "Radio channel characterisation for forest environments in the VHF and UHF frequency bands", *IEEE Proceedings 1999*

[26] L. Melin, M. Ronnlund and R. Angbratt, "Radio Wave Propagation. A Comparison Between 900 and 1800 MHz", *IEEE Proceedings,*

[27] European standard, ETSI 302 208, *European Telecommunications Standards Institute*, 2005

[28] D. W. Casbeer, R. W. Beard, T. W. McLain, L. Sai-Ming, R. K. Mehra, "Forest fire monitoring with multiple small UAVs", American Control Conference, 2005, 8-10 June 2005 Page(s):3530 - 3535 vol. 5

[29] Theodore S. Rappaport, "Wireless Communications", Prentice Hall PTR, IEEE Press, ISBN 0-7803-1167-1, 1996

Wireless communication system for a wide area sensor network

Magali Cortez and Jaime Sánchez

Department of Electronics and Telecommunications, Applied Physics Division
Center of Scientific Research and Superior Education of Ensenada
Km. 107 Tijuana-Ensenada Hwy. Baja California, México
{mcortez,jasan}@cicese.mx
http://www.cicese.mx

Abstract. In this paper, the design and implementation of a wireless communication system specific to wide area sensor network capable of establishing ad hoc connections is presented. First, the application to which this system is applied is explained, and then a brief description of the system is made. In this system two routing algorithms were implemented, one based on a static links routing table and the other is the M to 1 algorithm proposed by Lou W. in [1]. Tests were carried out with the designed system and the routing algorithms mentioned, accomplishing satisfactory communication between nodes in a test scenario for the first case.

Key words: Sensor network, wireless networks, wide area, microcontrollers, CR10X acquisition card

1 Introduction

Nowadays, wireless sensor networks represent one of the most innovative technologies. In general, this type of networks consists of a certain number of nodes distributed within an area. The nodes are provided with some type of sensor, and thus are used to collect and process data [2]. No matter what the application be, the sensor nodes that integrate this network have similar characteristics: small size, a transmitter-receiver module, a microprocessor and at least one sensor; limited in energy comsumption, transmitting power, memory and processing capacity [3]. These nodes communicate by forming an ad hoc network, in which data is being sent from a source node to a destination node through intermediate nodes. This way, the covered range of the network is extended. There are commercial wireless communications systems developed under the IEEE 802.15.4 standard for wireless sensor networks that offer personal area coverage, which are suited in applications where the nodes are distributed in a small area or where the density of sensors per unit area is large. The problem arises when a larger area is trying to get covered, at an urban or greater level, where the nodes are separated by long distances, and to form an ad hoc topology each node requires to transmit with enough power to reach to its neighbors, which

Please use the following format when citing this chapter:

Cortez, M., Sánchez, J., 2007, in IFIP International Federation for Information Processing, Volume 248, Wireless Sensor and Actor Networks, eds. L. Orozco-Barbosa, Olivares, T., Casado, R., Bermudez, A., (Boston: Springer), pp. 59-70.

is not possible with commercial wireless sensor network systems due to the fact that these transmit with very little power (some dBm). The Network of Deformations of the Mexicali Valley (RDEFVM) of the Deparment of Sismology at CICESE [4], presents a scenario where the sensor nodes are being separated by a few kilometers. The collection of data generated by each node is done manually by monthly visits which cause a delay in the analysis of the acquired data, and implies an economic and physical burden, specially on summer months when the Valley's high attainable temperature may reach from 40 to 50 C. A remote recollection of the data with a wireless network would avoid those problems, with a given benefit to the researchers in charge of the RDEFVM.

This report presents the design and implementation of a wireless communication system specific to a wide area sensor network capable of establishing ad hoc comunications. A graphical user interface necessary to the control and recollection of data, and a web page for the remote recollection of data via internet are developed. The rest of the document is organized as follows. In Section II, a description of the system designed in this project is given and the routing algorithms implemented are explained. In Section III, the test scenario is explained and the results obtained are presented later. Finally, in Section IV, the conclusions of this work are presented.

2   Structure of the system

This section describes the general architecture of the Wireless Sensor Network, the design methodology, the routing algorithm and the graphic interface developed for the user.

2.1   Wireless long range communication system.

In a wireless sensor network applications where the nodes are separated by long distances, each node is required to transmit with enough power to reach its neighboring nodes in order to create an ad hoc network. To provide wide area coverage to the RDEFVM, each node was equipped with a high power transmitter and a high gain antenna. To establish ad hoc communication within the network, each node had a microcontroller with a routing algorithm loaded previously in its memory. Because this system tries to give a solution to the problem of data collection that the RDEFVM represents, the characteristics of the nodes that integrate this network were considered in the design.

The RDEFVM has different types of sensors, but to begin, only displacement sensors (inclinometers) will be used. Therefore, each node in the network will have an inclinometer, a CR10X data acquisition card from Campbell Scientific and a 20W solar panel capable of powering the node. The sensor is connected to the card, which has the task of acquiring, processing and storing the data generated. The data acquisition card's output is through a serial communications port. Otherwise, to have access to the network from a remote device thorugh the internet, at least one of the nodes must be connected to it. A graphical user interface is necessary for the remote control of the network and the manual and automatic collection of the data captured by the nodes.

2.2   Network Architecture.

The architecture consists of the sensor nodes, an access node, a web page server and a remote terminal. The sensor nodes form the basic level of architecture of the system. A sensor node performs the network functions as well as sensing, since it processes and sends the data generated by itself and the rest of the sensors to the access node.

The access node is responsible for collecting and transmitting the information of the sensor nodes to a remote terminal via internet. The access node is a personal computer that communicates wirelessly to the sensor nodes and has specialized software for the recollection of data. The web page server hosts a web page necessary for the remote access to the network via itnernet. For simplicity, the server can be installed in the acces node. Lastly, the remote terminal can be any computer with internet access. Figure 1 shows the architecture described previously.

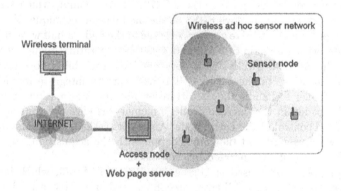

Fig. 1. System architecture

2.3   Design of the sensor node.

Originally, the RDEFVM's nodes were equipped with a sensor, a data acquisition card and a battery connected to a solar panel. The following paragraphs present the design of the electronic circuit that needs to be added to each node in order to establish wireless comunication. From this on, the electronic circuit added will be called the wireless interface.

Figure 2 shows the block diagram of the sensor node, name which is endowed to the original node of the RDEFVM once the wireless interface is added. The wireless interface can be divided into two blocks: one represents the wireless communication and the other the logical control.

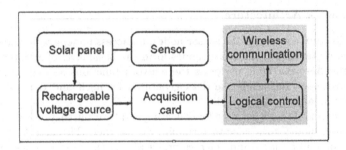

Fig. 2. Sensor node block diagram

Due to the fact that the nodes are separated by distances measured in kilo-meters, a high power card is required in order for the signal emitted by a node can reach some other neighboring node in the network. The wireless card by which the network is designed is the AC4790 of Aerocomm, which transmits with 1W of power and employs FHSS as the modulation technique [5].

On the other hand, to give greater coverage to the node, a half wave magnetic dipole antenna with a gain of 2 dBi and a 900Mhz operating frequency is used.

Since the AC4790 wireless card does not have the ability to establish ad hoc communications in the network, it is necessary to integrate in each node a microcontroller to handle the data transfer and link information. A program previously loaded in the microcontroller's memory handles the routing as well as the logical control of the entire node. Among its tasks is the communication with the wireless card and the comunication with the acquisition card for the recollection of the data generated by the sensors.

The microcontroller used is Microchip's PIC18LF542 [6], which has serial communications through a TX transmission line and an RX reception line. Since it is necessary to establish communication with the acquisition card as well as with the wireless card, a switch is required to choose between those two devices. The SN74CBTLV3257 from Texas Instruments is a digital switcher, better known as a MUX/DEMUX, from one to two lines of four bits, and is the device used for such purpose. Figure 3 shows a block diagram for the conection between those three devices and figure 4 shows a picture of the wireless interface prototype designed.

2.4   AdHoc communication.

The packet flow in an AdHoc network is from a source node to a destination node, using intermediate nodes as relays. This class of communication may be done by either broadcast or routing algorithms. In the broadcast mechanism, packets

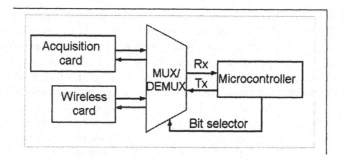

Fig. 3. Chip select function performed by the microcontroller.

are sent from a source node and are retransmitted by the neighbor nodes until they eventually reach their destination.

Fig. 4. The wireless interface prototype

There are two options for the broadcast mechanism: flooding and rumor broadcasting. In flooding, each node checks the arriving packets and if it happens to be the destination, it process the packet; otherwise it retransmits the packet to all its neighbors, as long as the maximum number of hopes has not been reached. This mechanism does not need to know the network topology, hence adding or deleting nodes to the network is an easy process. However, this mechanism has several problems, among them is the implosion problem, which means that duplicate packets are sent to the same node, and the superposition that occurs when two nodes send the same packet simultaneously.

Another broadcasting mechanism is called rumor, which is similar to flooding, with the difference that this one retransmits the packets to just one neighbor (which is chosen in a random way), which reduces the implosion problem [7].

The routing algorithms base their operation in a table that contains routes to reach all other nodes. A route is a sequence of hops to other nodes, needed to reach the destination node. Each node has its own routing table, depending of it location inside the network. The network topology must be known by all nodes, and it is necessary a route maintenance mechanism to update the routing tables in case of any change. An example of a routing algorithm with a route maintenance mechanism is the AODV [8], which was designed for Mobile AdHoc Networks where the network topology changes frequently.

In a network with few nodes, the broadcasting mechanisms are more adequate since the packets may reach the destination node with relatively few retransmissions. Otherwise, if a routing algorithm is employed, the power consumption may be optimized since the route to deliver the packets is known in advance, and just the necessary retransmissions are made; however, it needs memory space to store the routing tables. Since the RDEFVM nodes are fixed, and the geographical location of each node is known in advance, it was decided to implement a routing algorithm based on a static routes table; considering that the microcontroller chosen for implenting the AdHoc communication has enough memory to store a small routes table.

2.5   Routing algorithm.

The static routes table used for the network counts with three fields: destination node, next node and number of routes to the destination node. Once the destination node is known, the next node is looked up in the table so the package can reach its destination. The number of available routes is checked in order to get to the destination node, if there is more than one possible route, and the package cannot get to its destiny using the first route, then the remaining routes will be used.

Roughly, the microcontroller constantly checks the serial port, waiting for a packet to arrive. If the address of the node receiving the packet matches the packet destination address, the remaining of the header is processed, and the corresponding actions are executed on the acquisition card. In this case, an internal packet is constructed and sent to the acquisition card, asking for an answer to a command. Once the acquisition card performs the asked task, the microcontroller receives the answer and constructs a packet to be sent to the Access Node. The constructed packet includes the next node address for transmitting the packet to the destination node, that is obtained by consulting the routing tables.

The constructed packet is sent by the node through the wireless communications card, and waits for an Acknowledge (ACK) from the next node. Once the ACK is received, the microcontroller initializes all variables and registers used and waits for the next packet (from the wireless card) to be processed.

In case of receiving a negative ACK, the constructed packet will be retransmitted until a positive ACK be received or until the number of retransmissions is reached. The retransmission opportunities include the search for a new route, if the retransmission number has been reached for a particular route.

Whenever an alternate route to the destination node exists, which means that the number of routes field in the routing table is greater than one, the packet is sent again until acknowledged or until the retransmission number is depleted. If still the packet is not received correctly, the session is closed and the registers and variables are reinitialized, and the process remains waiting for a new packet. When the packet destination address does not match that of the node receiving it, the packet is retransmitted to the next node en route to the final destination. To avoid the possibility of remaining in a waiting state forever, a timer is initialized at the beginning of the interaction process of the microcontroller with the acquisition card. In case the timer expires (after 2 seconds), and no answer be received from the acquisition card, the card is considered damaged and the message error in the acquisition card is generated and sent to the access node. Once the error message is sent, the microcontroller choses the antenna signal and returns to the waiting loop, checking for the reception of a packet through the serial port, returning to a stable situation.

Packets used to implement the routing algorithm. The structure of the packets used for implementing the routing algorithm is shown in figure 5. Each packet contains the field name and the number of bytes for each field. Three of the packets use a common 7 bytes header, which is needed for the radio chip (AC4790) to configure some parameters for the payload transmission. The header contains the following information: packet header identifier (81H), payload size (128 byte maximum), session duration (time during which the communication among nodes is active), number of retransmissions (256 maximum), and MAC address of the next node.

Command type, code 0x01 (7 bytes header + 45 bytes)

81	Payload Data Length	Session count refresh	Transmit Retries	Next hop MAC	Message type	Destinatio n MAC	Command Type	Available bytes
1	1	1	1	3	1	3	1	40

Data type, code 0x02 (7 bytes header + 128 bytes)

81	Payload Data Length	Session count refresh	Transmit Retries	Next hop MAC	Message type	Destinatio n MAC	Data + MPTR
1	1	1	1	3	1	3	124

Error type, code 0x83 (7 bytes header + 4 bytes)

81	Payload Data Length	Session count refresh	Transmit Retries	Next hop MAC	Message type	Destinatio n MAC
1	1	1	1	3	1	3

ACK type, code 0x82 (4 bytes)

82	0x00: Failure, 0x01: Success	RSSI	RSSI*
1	1	1	1

Fig. 5. Packets used for implementing the routing algorithm in the system.

Inside the payload field is added another header, which is used by the microcontroller to identify four types of packets: Instruction, Data, Error, and ACK. The first byte value of this header identifies the type of packet and the next three bytes specify the destination address of the packet. Each packet may include other fields on its header, depending on the packet type.

The M-to-1 algorithm. A routing table is easy to implement and is suitable for scenarios where few fixed nodes are involved. However, if a node ceases to be a part of the network or if a new node is added (in other words, the topology of the network changes), an updating mechanism for the routing table is needed, otherwise the communication of the network would be hampered because the nodes would not count with valid routes; for this reason, an M to 1 algorithm proposed in [1] was used.

This algorithm takes advantage of the simple flooding by adding two mechanisms which, at the end of the flooding allow the nodes to know alternative routes to get to the base node, which is the node that initiates the flooding.

The M-to-1 algorithm bases its operation in the fact that at the end of a simple flooding, a tree is obtained, which is headed by the base node with branches formed by the closest nodes in which each branch is independent. This algorithm tries to unite the mentioned branches to provide more return routes for the base node to get to each node.

The message structure employed is type, mid, nid, bid, cst, route, where type is the type of message, in this case it's RPRI; mid is the number of the sequence of the updating route; nid is the message sending node's identifier; bid is the branch identifier of the node that sent the message that is the nid of the nearest node to the base node, cst is the cost of the route that could be the number of jumps; route is the path by which the package has traveled. For additional details regarding the M to 1 algorith, consult [1].

Integration of the M-to-1 algorithm to the designed system. The M-to-1 algorithm allows the nodes to have at least one route back to the base node, or one-way communications. In order to provide the system two-way communications, it is necessary to add another algorithm that allows the base node to know the route(s) to get to a determined node. The algorithm added for that purpose uses the following messages: RREQ and RREP.

RREQ is a routing request sent by the base node. RREP is the response to that petition that the destination node sends to the base node when it receives RREQ. RREQ is a broadcast message while the RREP is sent following a route discovered previously by using the M-to-1 algorithm.

The RREQ message counts with the fields Type, destination (broadcast), number of jumps, sequence number, destination node, previous node. The RREP message includes the fields Type, destination, sequence number, origin node, message sender node. When the z node receives a RREQ message, it updates the number of jumps and checks the destination node, if it is its neighbor it sends a RREQ placing in destination the address of the neighbor node and in previous node it's own ID. If the destination node is not its neighbor, then it resends the message, updating the number of jumps and placing it's own ID in the previous node field. When the RREQ reaches it's destination node, this node sends a RREP, in which destination is the base node, it's own ID in the origin node, and in the sending node it places it's own ID. This RREP is sent to the node which is indicated in the previous node field of the RREQ message being answered. When the z node receives a RREP message, it updates it's routing table with the information contained in the message (destination node = origin node; next jump = sending node). The z node places in the sending node field it's own ID, searches in the routing table a route to get to the base node (previously discovered with M-to-1 algorithm) and it sends a RREP message.

The process consists of flooding the network with the RPRI message. After that, the base node sends a RREQ message to know the route to get to the z node. The base node waits for a RREP message sent by the z node. After a t period, if the RREP message has not been received, a RREQ is sent again with a new sequence number; if the RREP message is received, a request of data is sent to the z node using the route contained in the RREP message. To ask for data from the acquisition card to the z node, a REQUEST message is employed.

2.6   Access node.

The access node controls the collection of data generated by the nodes in the network. This node consists of a personal computer that communicates wirelessly with the sensor nodes and has software designed for the collection of data from the nodes. It's installed in a location that has 120V AC power and internet acces available. To endow wireless communication to the access node, a multiport card was used included in the designer's kit by Aerocomm. That card includes an AC4790 card and communication options via RS232, RS485 and USB.

2.7   Graphical user interface for the control and collection of data and a Web page for remote access.

The program of the access node was made in LabView 7.1. The program sends the intructions necessary for the download of the data from the acquisition card, using packets that contain appropiate header to travel thorugh the wireless network to get to their destination. In the following paragraphs the program for control and collection of data is described.

The program performs the control funcions and the collection of data for the sensor network. The collection of data can be manual or automatic. In manual mode, the user has to select the desired node to connect, configures some parameters of the serial communication and begins to download. With automatic download, the user has to program beforehand a list of tasks. The list of tasks indicates the hour and the node number from which the data needs to be collected. In both cases, manual and automatic, at the end of each succesful download the data collected is being sent via email to the registered users of the system. Figure 6 shows the GUI developed.

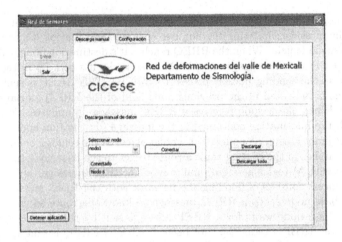

Fig. 6. Graphical user interface for the control and collection of data.

To have remote access to the wireless sensor network via internet, a web page as well as a web page server are needed. For the sake of simplicity, the web page server included in LabVIEW version 7.1 was used, which allows to publish web pages that contain Virtual Instruments. This way, the user can have access to the same program installed in the access node through a web page. The web page server is installed in the computer designated as the access node, so it is necessary for that computer to have a public IP address so that the users can access the web page.

3 Test scenario and results.

The system was tested making downloads of data from some of the sensor nodes which were placed inside the CICESE campus, where an ad hoc topology was employed and a GUI ran on a personal computer. The transmitting power of the AC4790 cards was 10 dBm and the maximum distance between the nodes was 300 m. Of course, the output power of the AC4790 can be set up to 1W to achieve greater coverage.

The AdHoc communication between the nodes, using the algorithm based on a static routing table was satisfactory, achieving 100 % of connections. Nevertheless, when testing the system with the M-to-1 algorithm and the forward communicating algorithm, problems caused by collisions between broadcast messages appeared. This problem is related to the AC4790 card, which cannot resolve the collisions generated by sending broadcast messages at the same time by different nodes. The solution to that problem was not addressed in this work, and is a pending task.

4   Conclusions

The wide area wireless communication system designed provides ad hoc communication using devices intended for point to point links. The results obtained were satisfactory, achieving multi jump communication in a testing network platform. In principle, the system designed tries to give a soultion to the problem of data collection imposed by the RDEFVM. However, the system developed can be used in any application that requires long range wireless communications and that employs the CR10X data acquisition card from Campbell Scientific.

Acknowledgments. The authors want to thank CONACYT, for their support for project No. 7410 "Deformaciones de la Corteza en el Valle de Mexicali", in charge of Dr. Ewa Glowacka, from which this work was originated.

References

1. Lou W.: An efficient N-to-1 multipath routing protocol in wireless sensor networks, 2nd IEEE International Conference on Mobile Ad-hoc and Sensor Systems, Washington DC. (2005).
2. Agrawal D. P., Biswas R., Jain N., Mukherjee A., Sekhar S. y Gupta A.: Sensor systems: state of the art and future challenges. En: Wu J. (ed.). Handbook on theoretical and algorithmic aspects of sensor, ad hoc wireless, and peer-to-peer network. Auerbach Publications, New York (2005) 318–345.
3. Haenggi M.: Opportunities and Challenges in Wireless Sensor Networks. En: Ilyas M. y Mahgoub I. (ed.). Handbook of sensor networks: compact wireless and wired sensing systems. CRC Press, Florida (2005) I–1.
4. RDEFVM: http://sismologia.cicese.mx/Redes/Rdefvm.html (2006)
5. Aerocomm: AC4790 900 MHz OEM transceivers. Users Manual, version 1.3 (2006).
6. Microchip: http://ww1.microchip.com/downloads/en/DeviceDoc/39631B.pdf (2006)
7. Heinzelman W.R., Kulik J. y Balakrishnan H.: Adaptive protocols for information dissemination in wireless sensor networks. 5th annual ACM/IEEE international conference on Mobile computing and networking (1999) 174–185.
8. Perkins C. E. y Royer E. M.: Ad hoc On-Demand Distance Vector Routing. 2nd IEEE Workshop on Mobile Computing Systems and Applications (1999) 90–100.

Anonymous Proactive Routing for Wireless Infrastructure Mesh Networks

Alireza A. Nezhad, Ali Miri, Dimitris Makrakis

University of Ottawa
800 King Edward Ave., Ottawa, Ontario, Canada
{nezhad, samiri, dimitris}@site.uottawa.ca

Luis Orozco Barbosa

Instituto de Investigación en Informática
Universidad de Castilla-La Mancha. Campus Universitario
s/n 02071 Albacete, SPAIN
lorozco@dsi.uclm.es

Abstract. An overlay routing protocol for infrastructure mesh networks is proposed that preserves user location privacy, source anonymity, destination anonymity and communication anonymity against an omni-present eavesdropper, when the underlying routing protocol is based on a proactive approach. A client only trusts its immediate access router. In order to receive packets, a client establishes a secret hop-by-hop virtual circuit between an arbitrary router, called its *Rendezvous Point (RP)* and its own access router, ahead of time. Packets destined for that client would be sent to RP first. To thwart content analysis attacks, we have used per-hop encryption. Authenticity and confidentiality of exchanged messages are also ensured using a public key infrastructure (PKI).

Keywords: Location Privacy, Anonymity, Ad hoc Routing, Mesh Networks

1 Introduction

Recent advances in wireless communications have presented the research community with new challenges in regards with user security. Two important aspects of security in mobile wireless networks from the users' perspective are *location privacy* and *anonymity.* In this article, location privacy means unlinkability between the location and the identity of a user. Anonymity in communications can be categorized as sender, receiver and relationship anonymity [10]. Relationship anonymity, sometimes called *communication anonymity,* means that a third party cannot identify both the sender and the receiver of a message or both ends of a certain connection (data flow). If the communications activities of a mobile device can be monitored, then the identity and

Please use the following format when citing this chapter:

Nezhad, A. A., Miri, A., Makrakis, D., 2007, in IFIP International Federation for Information Processing, Volume 248, Wireless Sensor and Actor Networks, eds. L. Orozco-Barbosa, Olivares, T., Casado, R., Bermudez, A., (Boston: Springer), pp. 71-82.

the movement patterns of the user of that device can be revealed, which violates the anonymity and location privacy rights of that user.

Obviously, there are many ways in which the privacy rights of a user may be violated including unauthorized access to the databases of context-aware applications containing location samples, user identification and locating at the time of association with the network, location-dependent temporary IP addresses and RF fingerprinting, to name a few. However, in this paper we are concerned with preventing locations and identities of communicating devices from becoming known to unauthorized entities as a by-product of inherent functions of routing protocols in multihop ad hoc networks.

Wireless devices are usually limited in terms of radio coverage. In order to make communication between two distant nodes in a wireless networks possible, cooperation of other nodes is necessary. This gives rise to what is referred to as *multihop* wireless communications. As the name suggests, the path between two end-nodes may traverse multiple intermediary nodes. Finding and establishing such a path is the important task of routing protocols. In wireless networks, there are two main classes of wireless routing protocols, usually referred to as *proactive* and *reactive* routing protocols. Despite their long successful history in wired networks, proactive routing protocols (e.g. OSPF and RIP) proved at first to be inefficient in wireless networks. This was mostly due to their large control overhead generated by periodic routing updates needed to keep nodes' routing tables correct at all times, in the face of frequent changes in topologies of wireless (especially mobile) networks. In the frequently changing topologies of ad hoc networks, these updates have to be broadcasted more often, which means more consumption of power and bandwidth. Another problem with proactive routing protocols was their memory requirement in order to store routing tables on each node containing routes to every possible destination. Because of these reasons, the new reactive routing protocols designed for ad hoc networks (e.g. DSR [1] and AODV [2]) proved to be more efficient and scalable than their proactive counterparts (e.g. DSDV [3]). These protocols only create routes when they are needed and discard them when they are no longer used. However, this behavior results in the so-called "*slow start*" problem, which introduces a path setup delay.

Because of the success of reactive routing protocols in ad hoc networks, almost all of the efforts in the field of anonymous routing for this kind of network were also focused on this class of routing protocols. ANODR [6] and MASK [7] are examples of these protocols. However, due to advances in radio technology on one hand and the introduction of improved proactive routing protocols on the other, the outlook is gradually changing. Bandwidths upwards of 100 Mbps are now available in wireless networks, which means larger amounts of routing updates can be accommodated. Also, mobile devices are nowadays equipped with much more memory. Several new proactive routing protocols for ad hoc networks (mostly based on link state routing) have been designed that reduce the amount of routing overhead significantly, via efficient dissemination techniques. Among these protocols, OLSR [4] and TBRPF [5] are now two of the three MANET RFCs in the area of routing. In this paper, we propose an anonymous routing protocol based on a proactive approach.

Regardless of being reactive or proactive, all of the early ad hoc routing protocols were designed without security in mind. One of the aspects of security that has been neglected in these routing protocols is *user location privacy*. Mobile users are not stationary and tracking of their movements through monitoring of their communications is a real concern. Another security-related shortcoming of regular ad hoc routing protocols is their lack of *communication anonymity*. Normally, the identities (IDs) of the source and the destination are contained in every data packet and hence known at the same time. Also, in reactive routing protocols, these IDs are present in route discovery messages.

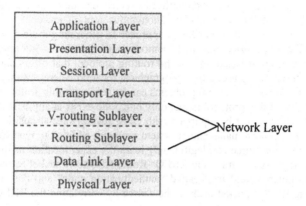

Fig. 1. V-routing protocol in the OSI reference model

Our *V-Routing* routing scheme is an overlay protocol that provides location privacy and communication anonymity to the end nodes of a data flow. As shown in Fig.1, it is an overlay protocol in the sense that it uses the services of an underlying proactive routing protocol in order to actually deliver packets. It allows the destination to establish a secret virtual circuit on top of the actual route in a way that its location is hidden even though it remains reachable.

In the next section, we provide a brief review of some related works. In section 3, we explain our network model. In Section 4, we outline our privacy objectives. In Section 5, our threat model is described. In Section 6, we provide a description of our proposed routing protocol. Finally, we conclude the paper with a summary.

2 Related Work

Recently, several protocols have been designed that add security including user anonymity and location privacy to regular ad hoc routing protocols. Furthermore, several new protocols have been introduced for this purpose that have been designed from scratch. However, in regards with location privacy and communication anonymity, virtually all of the efforts have been directed towards reactive routing protocols.

Kong and Hong presented ANODR [6], an identity-free anonymous routing protocol that uses route pseudonyms for each hop on the source-destination path, instead of node identities, in order to construct an end-to-end path. To reduce the cost and latency of its cryptographic onion approach, ANODR uses a novel technique called Trapdoor Boomerang Onion in the route discovery process that makes sure no local or global eavesdropper can learn the complete path. However, ANODR has several practical issues including its reliance on the existence of a global trapdoor that implies the source and destination have a pre-established shared secret, extra path setup delays at the intermediary nodes due to various symmetric and asymmetric cryptographic operations as well as a slow and overhead-expensive route repair mechanism.

MASK [7] is another identity-free anonymous routing protocol very similar to ANODR, except that it takes a different approach to generating route pseudonyms. In addition, it has high processing and memory requirements for intermediary nodes. AO2P [8] is a position-based on-demand routing protocol that offers communication anonymity. It delivers packets to a geographical location where destination has been reported to reside lately. Several problems can be seen with this protocol. For example, the premise of this protocol is that only one node exists at any particular position at any time. Nodes must be equipped with GPS, several special position servers are needed and the position management system produces additional overhead in the network. Besides, the source can legitimately learn the exact location of the destination node, eavesdroppers can trace the RREQ (Route REQuest) packet to the destination using an un-mutable field in it called "authentication code" and they can trace the RREP (Route REPly) packet back to the source by correlating the RREQ and RREP packets.

AnonDSR [9], which is an anonymous routing protocol based on DSR prevents a data packet from being traced back to the sender or receiver, but this protection is limited to the data transmission phase. In the route acquisition phase, similar to regular DSR, the identities of the two end nodes and all the intermediary nodes are transmitted as clear text. Moreover, the RREQ packet carries a temporary public key that is fixed across all the hops between the source and the destination. An omni-present adversary (a global adversary who can monitor all transmissions) can use this field to trace the packet back to the source and the destination.

3   Network Model

One of our main assumptions in this paper is that at any given time there are a number of legitimate member nodes in the network, which do not require location privacy and anonymity. Therefore, these nodes participate fully in the routing process i.e. they identify themselves to their neighbors, collect and broadcast their neighborhood information (about neighbors that authorize it) to other nodes by way of routing updates and forward packets for other nodes according to their own routing tables. We refer to these nodes as *routers* or *access points*. On the other hand, some of the nodes in the network may like to hide their locations and their movements as much as possible. We refer to these nodes as *ordinary nodes* or *clients*. If a client does not need location privacy, it may broadcast its neighborhood information but it will still identify itself as a client meaning it would not act as a router[1]. This way, it can avoid the costs

---

[1] We will explain later that this design is tailored towards 1-hop clustered architectures, such as infrastructure mesh networks. Another version of our protocol, not described here, is appli-

associated with the location privacy protocol. In other words, our protocols give the client the option to dynamically decide whether it wants to remain hidden and pay the cost of privacy or prefers to bypass our security mechanisms and reveal its location. Every wireless device is identified by a unique location-independent network identifier (ID) such as its real permanent IP address. We call a neighboring router of a client with which the client associates an *Access Router* of that client. For simplicity, in this paper, we consider only one access router per ordinary node. The access router of the source node (S) in a connection is denoted by AS and the access router of the destination (D) is denoted by AD. An access router does not advertise its membership information. A client connects to its access router on the link layer as in 802.11 without specifying its own MAC address. Instead, it uses the secret link layer key that it shares with its access router to hide its ID.

Clients who do wish to keep their locations hidden refrain from advertising their locations and neighborhood information. This network model is essentially a clustered architecture with access points as clusterheads, which are directly connected to their clients. An example, which is more akin to our assumed network structure, is an *infrastructure mesh* network in which clients do not participate in routing. At this time, we are not considering *client-mesh* networks where clients also help in routing packets. Mesh technology is becoming increasingly popular with applications in consumer, small business, metropolitan, and military situations, to name a few. A mesh network is typically a network of WLANs where only one or a few access points are directly connected to the wired world and act as gateways to the public networks. Other access points use multihop routing in order to access one of these gateways or any other access point, effectively forming a wireless virtual backbone. An example of this kind of network architecture is a public network of WLAN hotspots e.g. a Wireless Internet Service Provider (WISP). For instance, recently, Toronto Hydro Telecom Inc. in Canada turned the whole Toronto downtown into a large WiFi zone [11]. In this network, many access points are deployed throughout a very large area forming a backbone that uses multihop routing to connect mobile users to the wired Internet. Another example is known as Wireless Community Networks, a confederation of WLANs usually meant to provide free Internet access to users. A long list of such networks can be found in [12]. Mesh networks are being widely used to easily and cost-effectively help municipalities, counties and organizations like departments of transportation overcome the challenges of rolling out fixed and mobile wireless data networks. It enables vehicles, mobile devices and individuals to instantly and securely connect directly with each other and to the public telephone network, the Internet and private networks for access to voice, video and data services. In such an environment, it is reasonable to assume that users of the system would not be happy to know that the network operator and all other users can potentially take advantage of the weaknesses in the ad hoc routing protocol to track all their movements. The two main proposals for mesh networking namely SEEMesh and Wi-Mesh have been merged to form a starting point for the 802.11s [14] extension to WiFi standard.

---

cable to a general mesh network as well as 4G-model ad hoc networks in which user devices belong to multihop clusters and participate in packet forwarding within their cluster. An example of this kind of networks is a multihop WiFi hot-spot.

The mesh architecture is also being considered by the industry to be used with Wi-Max (IEEE 802.16) technology, instead of the traditional point-to-multipoint configuration [15], [16].

4   Privacy Objectives

The design of our protocol depends greatly on our trust model e.g. where the destination's trust lies. In this article, we assume that a client trusts at least one router in its neighborhood (its access router) with its location and identifies itself to it in order to be able to receive packets. This is in accordance with the community network model and an ad hoc kind of network. For example, in a community network consisting of houses, a client may fully trust its access router because he/she is either the owner or a guest in the house. On the other hand, in the WISP model, a client may only trust its home domain while roaming. Another version of V-routing, not discussed here, is designed for that scenario. There may be any number of reasons for lack of trust in other routers including their vulnerability to intrusion, the so-called "big brother" problem and opportunistic public network operators especially when the routes pass through foreign domains.

The location of a client is considered to be known if its access router is known. The locations of the source and the destination must only be known to their own respective access routers. Communication anonymity has to be supported with regards to peers as well as the network. In other words, only the source and the destination must know that they are communicating. We provide this feature by ensuring sender anonymity. Only the destination of a packet can know who is the original sender of the packet. Receiver anonymity is provided with regards to all third parties except one router on the source-destination path, called a rendezvous point for that connection, as will be explained later.

5   Adversary Model

We assume the existence of an omni-present adversary sometimes called a *global adversary* that can monitor all transmissions throughout the network. The adversary is assumed to be very strong in terms of processing power and storage capacity but it cannot break the cryptographic measures used by legitimate nodes in bounded time. It launches only passive attacks (eavesdropping) not active attacks such as DoS (Denial of Service), impersonation, message modification, man-in-the-middle, etc. These attacks can be discovered and the culprit can be identified using intrusion detection techniques. The kind of adversary that we have assumed wishes to remain undetected until it gathers its desired information.

We assume that the adversary is capable of performing traffic analysis. Attackers use traffic analysis techniques including content analysis and timing analysis to gain meaningful knowledge about the data flows, even though the data itself may be encrypted. They can exploit things such as packet length, packet headers, timing of transmissions on successive links, traffic patterns and so on to infer valuable conclusions about communications in the network. While we do not concern ourselves with traffic analysis too much, we do address one aspect of *content analysis* that is related to routing. A global adversary can match certain fields in packets (that either do not change or change in a predictable fashion) across successive links in order to trace them back to the source and/or the destination. For example, an RREQ packet in AODV can be traced using its sequence number. To resist this kind of threat, all packets must look completely independent of each other on different links, a notion sometimes referred to as one-time packets. Two methods are in common use for achieving this effect; one is to use onion routing where a new layer of encryption is applied to the packet every time it is relayed, thus changing how it looks on consecutive links. Usually, the source node encrypts data with the key that it shares with the destination or with the destination's public key. The next hop forwarder, encrypts the received packet again for the destination, so on and so forth. The destination, applies sequential decryption in the reverse order until it recovers the original data. This method requires either a PKI infrastructure or pre-established shared keys between the destination and all the intermediary nodes. The second method is per-hop encryption/decryption (re-encryption) between neighboring nodes. For example, in IEEE 802.11, the WEP (Wired Equivalent Privacy) key option allows neighboring nodes to share a link layer secret key. In our protocol, we have used the second method to prevent content analysis attacks.

6 Proposed Routing Protocol

In the spirit of taking the proactive approach to routing, in order to avoid the path setup latency of reactive protocols, the V-Routing protocol works in a proactive manner as well. It consists of two parts; *path establishment* and *data transmission*. Before a client can receive data packets, it has to go through the path establishment phase and set up at least one secret route towards itself, starting at an arbitrary router. In other words, the main part of routing is in the control of the destination and that is the secret behind destination location privacy in V-routing.

Suppose that a source node S wishes to establish a connection to a destination node D. If D is a router, its location in the network is known because it broadcasts routing advertisements. However, if it is an ordinary node, our underlying table-driven routing protocols cannot find a path to reach it because it hides its location by refraining from broadcasting routing information.   S too, may be a router or an ordinary node. If it is an ordinary node, the V-routing protocol allows it to hide its location as well.

6.1 Path Establishment Phase

When an ordinary node D joins the network, it looks at the current network topology obtained from periodic routing update messages broadcasted by routers. From this information, it chooses one of the reachable routers anywhere in the network, to be its *Rendezvous Point* (RP). This router will be used as a transient destination for any packet destined for D. Therefore, any future source node S trying to send a data packet to D will send that packet to RP first, creating a triangular path, as shown in Fig.2. We denote the first leg of this triangular path with S-RP and the second leg with RP-D. Having a global view of the network topology, D is able to calculate a secret route from RP to itself according to a policy of its choice such as the shortest path[2]. However, for security reasons this path must be as unpredictable as possible. In fact, this information may be readily available to D from the routing updates of RP. In other words, depending on the routing protocol, these updates may specify the path from RP to AD.

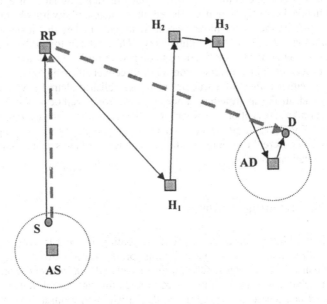

Fig. 2. Triangular Path in V-routing

D follows its own local policies to select RP. Reachability, distance and trust may be some of the criteria. Specifically, one factor that affects this decision is whether D can securely communicate with this router. D must either be able to obtain a signed public key for its RP (if public key infrastructure is used) or it should share a secret key with it (if symmetric cryptography is used). D uses unicast *forward_req* (for-warding request) messages in order to set up its second leg path. It secretly sends several such messages to a few selected routers of its choice, located along the RP-D leg, beginning with RP and ending with AD, in effect establishing a virtual circuit. We call these nodes *Virtual Hop Routers (VHR)* because each consecutive pair of them may or may not be physically one hop apart but will be able to reach each other,

---

[2] If D is not computationally powerful enough, AD may perform the path establishment on D's behalf but it should not identify itself in forwarding request messages.

using the underlying routing protocol. The format of a forward_req message is shown below. In this paper, $E_x(M)$ denotes an encryption of a message M for an entity x.

$$<forward\_req, RP\_elect, E_{VHR}(2^{nd}\text{-}leg\text{-}id, first\_VHR, D, next\_VHR)>$$

This message is delivered by the underlying routing protocol to the recipient VHR. next_VHR is the ID of the VHR downstream (towards D) from the recipient VHR. In order to prevent content analysis attacks, this message is re-encrypted at each hop. The boolean value RP_elect indicates to the recipient router whether or not it has been selected as a rendezvous point by D. 2nd-leg-id is a global identifier chosen by D that uniquely determines this 2nd-leg path and has the same format as IDs[3]. The Boolean field first_VHR will be explained later. D may obtain the authentic public keys of VHRs from a certificate authority or they may advertise their signed public keys in their routing update messages. Alternatively, if D shares a secret key with the VHR, they can use symmetric cryptography instead of PKI to communicate.

D sends its first forward_req message to RP asking it to forward any packets destined for D to a next_VHR, which we denote by H_1. If RP is willing to act as a rendezvous point for D, it broadcasts this decision in the network using an I_AM_RP packet. This packet is not encrypted and is readable by every one. Therefore, any node wishing to communicate with D will know it must send its packets to RP first. The I_AM_RP packet is shown below:

$$< I\_AM\_RP, RP, D, next\_VHR\_OK >$$

Due to the dynamic nature of ad hoc networks, there is a chance that the routing tables of D and RP may not match. Therefore, RP may not actually be able to reach H_1 even though it may be willing to act as D's rendezvous point. In the event of that happening, RP will set the Boolean field next_VHR_OK in I_AM_RP to FALSE and proceed to advertising its current routing information. After acquiring the correct routing information regarding RP, D chooses another router as H_1 and sends a new forward_req message to RP. If RP can reach the new H_1, it will set next_VHR_OK in I_AM_RP to TRUE.

Once RP is established, D sends unicast forward_req messages to H_1 and the other VHRs on the RP-D path. In each of these messages, D specifies the next virtual hop router for the recipient and sets the parameter RP_elect to FALSE. Of course, AD knows that it must forward packets destined for D directly to it on the link layer. This way, each VHR knows only its next VHR in order to reach D. Other nodes and eavesdroppers only know that D is using RP as its rendezvous point. Only D knows the entire RP-D leg. D can omit its ID from these messages, which enhances receiver anonymity.

If RP is not willing to act as the rendezvous point for D, it will refrain from broadcasting an I_AM_RP packet. Therefore, D can interpret the receipt of such a packet as an ACK (acknowledgement) from RP and its absence as RP's unwillingness or a NACK (negative acknowledgement), in which case it may try RP again (in case the previous message was lost.) or choose another router and repeat the same process.

---

[3] Hash functions can be used to generate identifiers that are globally unique. IETF RFC 4122 defines a namespace for globally unique identifiers.

Making RP, instead of D itself, responsible for disseminating this information in the network has two benefits. First, it prevents a nearby eavesdropper from locating D. Note that D's ID is not encrypted in the I_AM_RP packet because we want every node (potential source nodes for D in future) to learn the fact that D is using RP as its rendezvous point. Secondly, it reduces the overhead of the protocol because it lets RP to implicitly send an ACK to D (piggybacking). Thirdly, RP is able to take advantage of advanced flooding techniques already in use by the underlying routing protocol such as the "multipoint relay" method of OLSR to further reduce the overhead. D (or AD) needs to ensure that its 2nd-leg path(s) is always connected. Therefore, it must receive a notification from the underlying routing protocol when a change in the network topology is detected.

6.1.1 Acknowledging forward_req Messages

As was explained, RP acknowledges a forward_req message with an I_AM_RP packet. Other VHRs use a different mechanism for this purpose that carries a cumulative ACK message from all VHRs back to D. To enable this mechanism, we have to include a Boolean field *first_VHR* in the forward_req message, which is only set to TRUE for H_1. After receiving a forward_req message, H_1 assembles an ACK packet, which contains a long fixed length randomly filled data structure called *magazine*. We have chosen this name for this field because of its similarity to the magazine of a machine gun where the bullets are pushed in on top of each other and are later released in the reverse order. Another analogy for this field is a First In last Out memory stack. The format of this ACK message is shown below:

$$<ACK, E_{next\_VHR} (2^{nd}\text{-}leg\text{-}id), magazine >$$

H_1 pushes its ID (encrypted for D) at one end of the magazine and sends the message to H_2. H_2 and every other VHR on the path down to and including AD do the same. If VHRs use D's public key, each of them must first append a nonce (a one-time random number) to its own ID. Otherwise, a compromised intermediate router could systematically try each ID in the network encrypted with all the available public keys to eventually uncover each address in the magazine. A compromised VHR would be in a better position to mount this attack because it knows it only needs to try the public key of D. When a nonce is used, only D, using its private key, can uncover each filed in the magazine. Delivering an ACK packet from one VHR to the next one is the responsibility of the underlying routing protocol. Each VHR determines its next VHR based on 2nd-leg-id. Every VHR decrypts this field and then re-encrypts it with the public key of its next VHR. Per-hop re-encryption at the link layer is applied to the whole packet in order to prevent content analysis resulting in uncovering D's location.

If the ACK message traverses every VHR and successfully arrives at AD and is then forwarded to D, D will know that the second leg is complete. However, if some H_i is unable to reach H_{i+1} it will broadcast the so far accumulated ACK message. H_i encrypts the 2nd-leg-id for D. On the other hand, after waiting for a specified amount of time, H_{i+1} issues an ACK message if it does not receive one from a previous VHR. At the end, D may end up with a few contiguous segments with gaps between each two. At that point, for every *hanging* H_i (a VHR unable to reach its next VHR) D selects a new H_{i+1} (reachable by H_i and able to reach the old H_{i+1}) and sends a new forward_req message to H_i. D also sends a forward_req message to the new H_{i+1} specifying the old H_{i+1} as its next virtual hop router. By setting the *fisrt_VHR* to TRUE, D instructs the first hanging VHR in the chain to originate an ACK message and the process continues like the first time.

6.1.2 Data Transmission Phase

When a source node S wishes to send a packet to D, it sends it to RP as D's rendezvous point. This data packet is formed as:

$$<RP, E_D(S, nonce), E_{RP}(D, nonce), payload>$$

The main purpose of nonce is similar to what was explained in the path establishment phase. RP understands that it must forward the packet to D, the final destination. It replaces D's ID in the packet with the appropriate 2nd-leg-id encrypted for H_1 and forwards the packet to it. This field is re-encrypted at each VHR for the next VHR. In order to prevent content analysis using the un-mutating fields of source ID and payload, they are also re-encrypted at every VHR. S is also encrypted for the destination thus the communication is anonymous to all other nodes even the two access routers AS and AD. A data packet forwarded from H_{i-1} ($i > 0$ to include RP as well) to H_i looks like:

$$<H_i, E_{Hi}(E_D(S, nonce)), E_{Hi}(2^{nd}\text{-}leg\text{-}id), E_{Hi}(payload) >$$

Every VHR determines its next VHR based on the 2nd-leg-id and modifies the packet accordingly. At the last hop, the payload and $E_D(S)$ are forwarded by AD to D on the link layer using their shared secret key. These indirect packets are transported using regular IP protocols, e.g. according to the IP encapsulation specifications of the IETF RFC 2003 [13].

Summary

In this paper, we proposed a protocol for supporting user location privacy, user anonymity and communication anonymity in wireless infrastructure mesh ad hoc networks, where a client trusts its access router. We applied our general concept of *destination-controlled routing* to design one version of our V-routing protocol for this kind of network. To the best of our knowledge, V-routing is the only anonymous ad hoc routing protocol based on a proactive approach. Nevertheless, in our protocol, a user node does not advertise its neighborhood information and its location. Instead, it secretly and in advance, establishes a path towards itself starting at a transient destination (its rendezvous point), which receives packets destined for that user and then forwards them to it along the secret path. Source location privacy, source anonymity and communication anonymity are ensured by disclosing the identity of the source only to the destination. The identity of the destination of a packet is revealed only to its rendezvous point.

References

1. David B. Johnson, David A. Maltz, and Josh Broch: DSR: The Dynamic Source Routing Protocol for Multi-Hop Wireless Ad Hoc Networks, in Ad Hoc Networking, Chapter 5, pp.139-172, Addison-Wesley, 2001.
2. Charles E. Perkins, Elizabeth M. Belding-Royer, and Samir Das, Ad Hoc On Demand Distance Vector (AODV) Routing., IETF RFC 3561.
3. Perkins and P. Bhagwat, Routing over Multihop Wireless Network of Mobile Computers, in Mobile Computing, edited by Tomasz Imielinski and Henry F. Korth, Chapter 6, pp. 183-206, Kluwer Adademic Publishers, 1996.
4. T. Clausen, P. Jacquet, A. Laoiti, P. Minet, P. Muhlethaler, A. Qayyum, L. Viennot, Optimized Link State Routing Protocol, IETF RFC 3626, October 2003.
5. R. Ogier, F. Templin, M. Lewis, Topology Dissemination Based on Reverse-Path Forwarding (TBRPF), IETF RFC 3684, February 2004.
6. J. Kong, Anonymous and Untraceable Communications in Mobile Wireless Networks, PhD thesis, University of California, Los Angeles, June 2004.
7. Y. Zhang, W. Liu, and W. Lou, Anonymous Communications in Mobile Ad Hoc Networks, IEEE INFOCOM, Miami, FL, March 2005.
8. Xiaoxin Wu and Bharat Bhargava, AO2P: Ad Hoc On-Demand Position-Based Private Routing Protocol, IEEE Transactions on Mobile Computing, vol.4, no.4, July/August 2005.
9. Ronggong Song, Larry Korba, George Yee, *AnonDSR: Efficient Anonymous Dynamic Source Routing for Mobile Ad-Hoc Networks*, SASN'05, November 7, 2005, Alexandria, Virginia, USA.
10. A. Pfitzmann and M. Kohntopp. Anonymity, Unobservability and Pseudonymity -- A Proposal for Terminology, in Hannes Federath(Ed), Designing Privacy Enhancing Technologies, Lecture Notes in Computer Science, LNCS 2009, pp.1-9, Springer-Verlag, 2001.
11. http://michaelocc.com/2006/09/downtown-toronto-hydro-wifi-grid.html
12. http://en.wikipedia.org/wiki/List_of_wireless_community_networks_by_region
13. C. Perkins, IP Encapsulation within IP, October 1996, http://www.ietf.org/rfc/rfc2003.txt
14. http://www.ieee802.org/11/PARs/11-04-0054-02-0mes-par-ieee-802-11-ess-mesh.doc
15. http://www.wi-fiplanet.com/news/article.php/3549846
16. http://www.wi-fiplanet.com/news/article.php/3553326

Destination Controlled Anonymous Routing in Resource Constrained Multihop Wireless Sensor Networks

Alireza A. Nezhad[1], Dimitris Makrakis[1], Ali Miri[1]

[1] University of Ottawa, Ottawa, Ontario, Canada
{nezhad, dimitris, samiri}@site.uottawa.ca

Abstract. In this paper, a routing protocol is proposed that provides location privacy for the source and the destination as well as user anonymity and unlinkability in multihop wireless sensor networks. The sink is assumed to be computationally powerful and responsible for all routing decisions. It assigns incoming and outgoing labels to nodes in the uplink and downlink directions. Each node is only aware of its own labels and only forwards packets whose labels match either its downlink or uplink incoming label. Moreover, in order to prevent packet tracing by a global eavesdropper, layered cryptography is used in both directions to make a packet look randomly different on different links. However, due to the node capability limitations, only symmetric cryptography is used.

Keywords: Wireless Sensor Networks, Location Privacy, Anonymity, Routing

1 Introduction

Wireless sensors are characterized by limitations on batteries, computational capabilities (CPU power and memory), as well as communication capabilities such as bandwidth, transmission power and receiver sensitivity. In the early days of wireless sensor networks (WSN), the issue of network security was given second priority as the technology struggled to meet these strict and diverse constraints. It is only recently that a flurry of activities has been seen in some areas of wireless sensor networking including lightweight cryptography, secure routing and intrusion detection. However, *anonymous routing* that covers areas such as *node anonymity, node location privacy, untraceability and unlinkability* is yet to be adequately explored. These issues are of utmost interest in military, homeland security, and law enforcement but are also becoming progressively essential to many civilian applications.

Fig.1 demonstrates the importance of location privacy for the source and the destination in wireless communications. In this example, a criminal is interested in finding a valuable object (source) by using the beacon signals that it emits (similar to asset monitoring applications). It is also interested in identifying the locations of police patrols (destination, also called *sink* in WSN) in order to avoid or even eliminate them.

An anonymous routing protocol for WSN must prevent an adversary from finding the locations of the source and the sink. The adversary may exploit information in the packet headers or perform packet tracing. The former problem can be addressed

Please use the following format when citing this chapter:

Nezhad, A. A., Makrakis, D., Miri, A., 2007, in IFIP International Federation for Information Processing, Volume 248, Wireless Sensor and Actor Networks, eds. L. Orozco-Barbosa, Olivares, T., Casado, R., Bermudez, A., (Boston: Springer), pp. 83-94.

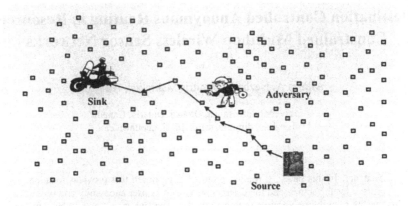

Fig. 1. Importance of location privacy for source and destination in a WSN

using encrypted packet headers or identity-free routing, while the latter must be addressed by untraceable routing schemes. According to Pfitzmann and Kohntopp [11], anonymity is the state of being un-identifiable within a set of objects; the *anonymity set*. Untraceability refers to the inability of an adversary in tracing individual data flows back to their origins or destinations. Unlinkability means preventing an adversary from learning the identities of the source and the destination at the same time.

Regular routing protocols designed for *multihop* wireless sensor networks [7] are vulnerable to *location privacy* violations. In order to achieve the highest possible efficiency, these protocols use methods that unfortunately offer a lot of help to an attacker. These methods are usually based on disclosing network topology, especially the location of the *sink*. Usually, the sink emits periodic beacon signals that are flooded in the network to let each sensor calculate the best (e.g. the shortest) path to the sink. Moreover, during data transmission, an adversary can trace data packets back to their source and/or their destination over consecutive links. This can be done using signal detection techniques such as triangulation, trilateration, angle of arrival, signal strength and so on in each locality to find the immediate sender of a packet.

Encryption is undoubtedly one of the most effective means of providing security services. However, encryption alone is not enough for ensuring anonymous communications. Adversaries can use *traffic analysis* techniques such as content analysis and timing analysis to obtain valuable information including the locations of the source and the destination of each packet as well as the identities of the nodes involved. Traditionally, anonymous communication schemes such as Mixes [12] have been used to protect networks against such attacks. Unfortunately, these solutions, even those versions developed for MANET [8], are not applicable to resource-constrained sensor networks. Lately, several anonymous routing protocols [1]-[6] have been introduced for wireless ad hoc sensor networks, some of which attempt to defend against this kind of attack. In the next section, we briefly overview some of these proposals and point out their shortcomings.

In this paper, we introduce a novel routing scheme for sensor networks based on label switching that supports traffic untraceability in order to hide the locations of a source sensor and the sink. It is also an identity-free routing protocol and as such offers anonymity to the source and the sink. The identity-free routing and untraceability ensure unlinkability even when multiple sinks use the same sensor network.

The rest of this paper is organized as follows: in the next section, we review some related works. In section 3, we describe our network model. In section 4, we explain our threat model and privacy objectives. In section 5, we present our routing protocol. Finally, we will conclude with a summary.

2 Related Work

Phantom routing [1] is one of the earliest works that clearly addresses the problem of source location privacy in wireless sensor networks. It considers that movements of an object are continuously monitored by a network of stationary sensors and that an adversary tries to locate the object by tracing the stream of packets flowing from its point of presence towards the sink. In order to mislead the adversary, phantom routing first sends each packet from the source randomly to an intermediary node called a *phantom source* located a number of hops away from the real source. The phantom source then either floods the received packet in the network or unicasts it to the sink. Obviously, the flooding version of this protocol does not reveal the location of the destination but it has the disadvantages of flooding, including excessive energy and bandwidth consumption. On the other hand, its single-path version cannot hide the location of the sink as it uses regular single-path routing schemes that require all sensors to know where in the network the sink resides.

Hong et al. [2] introduce two algorithms to protect sensor networks against timing analysis attacks. Timing analysis is a kind of traffic analysis that exploits timing correlation of transmissions from neighboring nodes to correlate packets on successive links and therefore uncover end-to-end traffic flows between the source and the destination of those packets. Most of the traditional countermeasures against this kind of attack, such as packet reordering and decoy traffic, are not suitable for the resource-constrained sensor networks. The authors base their algorithms on adding random delays to packet retransmissions at each forwarding node, in an attempt to obfuscate the temporal relationship among packet transmissions in a neighborhood. This paper is not concerned with the actual routing of packets. In other words, it does not specify how a path for a packet is found towards the sink.

Another kind of traffic analysis attack tries to gain sensitive information about end-to-end data flows by monitoring changes in link-level traffic patterns. The heuristic algorithm proposed in [3] tries to prevent this type of attack by routing end-to-end data flows in a way that keeps the global view of link-level traffic patterns on all the links across the network as invariable as possible. In fact, when the real traffic pattern in the network changes, this algorithm reroutes traffic flows in a manner that the overall network traffic patterns remain unchanged. However, this paper does not discuss implementation issues of this algorithm and their privacy implications. This algorithm requires complete knowledge of link-level and end-to-end flows throughout the network. In a centralized implementation of this algorithm for sensor networks, the sink would be the sensible choice as the entity performing the algorithm. In that case, protocols are needed for conveying necessary information from the sensors to the sink, as well as procedures for setting up the right link-level connections by the sink, all in a secure and private manner. A distributed implementation is more challenging, especially from a security standpoint. It would mean that global knowledge of data flows and network topology must be available to all sensors.

The focus of the work reported in [4] is protection against intrusion attacks and traffic analysis attacks that aim at isolating or locating the sink. It employs two main techniques to prevent such attacks. First, it increases tolerance against sink isolation by using redundant sinks and develops secure multipath–capable path setup mechanisms, so that sensors can report their data to multiple sinks. However, this path setup mechanism is based on sink-originated beacon signals, which reveal the location and identity of the sink. It also needs all nodes to identify their neighbors, which may also be a privacy concern. The second countermeasure offered in this paper is the use of anti-timing analysis techniques that prevent an attacker from tracking packets back to the sink by monitoring the transmission times of a sensor and its parent(s). It achieves this goal by unifying the transmission rates of all nodes, using delays and dummy packets, when necessary. This technique suffers from waste of energy and bandwidth, data loss due to buffer overflows and latency due to random delays.

Deng et al. [5] use four traffic randomization techniques to protect against traffic analysis attacks that aim at locating the sink in a WSN. They try to increase randomness in traffic patterns, in order to confuse an attacker who exploits pronounced traffic patterns due to fixed-path routing protocols. Traffic analysis attacks of the "rate monitoring" and "time correlation" types have been considered in this paper. The first technique proposed in this work is a per packet random multiple-path forwarding scheme used at each node. This scheme is based on the common topology discovery method of propagating beacons from the sink, which as mentioned before, reveals the location of the sink. The second technique is a controlled random walk that uses a probabilistic forwarding scheme at each node to determine the next hop according to a uniform probability distribution, in order to protect against rate monitoring attacks. The last two techniques use fake paths and fake hotspots that are based on decoy traffic, which as mentioned before, consumes energy and bandwidth as well as degrades performance in terms of packet delivery success ratio and latency due to increased rate of collisions.

Olariu et al. [6] hide the identities and locations of the source and the destination in a wireless sensor network, as well as provide traffic untraceability by imposing an anonymous virtual infrastructure over the physical infrastructure. However, their simple routing protocol is vulnerable to packet tracing. As mentioned before, several anonymity protocols have been developed for MANET, such as ANODR[8] and MASK[9], but they are not suitable for sensor networks due to their resource limitations, especially in terms of processing power. Mist Routing [10] can also provide source and destination location privacy, but it assumes a fixed infrastructure of special routers with a pre-determined logical hierarchy overlaying the physical network. Such assumptions are not typically applicable to wireless sensor networks mainly because sensors are usually deployed in an ad hoc manner.

3   Network Model

We envision a network of many (hundreds) small wireless sensors that are deployed in a large geographical area. We also assume that a single, considerably more powerful node, called the *sink,* exists in the network that controls the sensors and collects sensory data from them. Because of their limited transmission range, sensors send their data to the sink using multihop communication. Sensors generate data independently and relay packets received from their neighbors without deterministic knowledge of their arrival times. In other words, a node can always detect packet arrivals from its neighbors[1] and is ready to retransmit them according to a certain policy. Data packets generated by sensors always travel upstream (uplink) towards the sink and are never destined for any other node. Control packets such as routing updates from sensors also travel upstream, destined for the sink, while control packets from the sink such as routing instructions, cryptographic assignments, acknowledgments and so on, travel downstream (downlink).

Our approach to ensuring anonymity and location privacy for the destination in a network is to avoid disclosing information about the destination identity and its location as much as possible. To this end, we have made the sink entirely in charge of how packet routing should be done. That is why we describe our proposed routing protocol as being destination-controlled. Through a topology discovery process, described later, the sink obtains a global view of the network topology, but no other entity knows which node is the sink. We will also explain how the sink can securely and anonymously send instructions to each node in a manner that ultimately all packets can find their ways to the sink, without the privy of any internal or external entity other than the sink.

4   Threat Model and Privacy Objectives

We intend to resist an omni-present adversary sometimes called a *global eavesdropper.* This kind of adversary can monitor all transmissions happening anywhere in the network. It can theoretically follow one single packet to its destination. It can also detect the source of a packet as soon as it is generated. Therefore, in order to have an acceptable degree of protection against this kind of adversary, we need to have a sufficient number of active sources in the network, making the process of distinguishing a specific flow among many flows difficult. The greater the number of data flows (anonymity set) the higher the degree of anonymity will be.[2] This condition is intrinsically satisfied in many types of sensor networks such as inventory tracking, habitat monitoring and environment monitoring applications. Also, given a specific packet and the entire transmission history of the network, we would like to make the task of identifying the sender of the packet as difficult as possible for an eavesdropper. Finally, and most importantly, we wish to hide the location and the identity of the sink. Our

---

[1] Some energy-efficient MAC protocols allow a node to detect packets even in the idle mode.

[2] As an extreme case, consider that only one data flow exists in the network. Regardless of what routing mechanism is used, a global adversary can monitor all packet transmissions. In fact, it can detect the source as soon as it generates a packet.

protocol works on the network layer but it does not preclude the use of node level mechanisms such as anti-timing analysis techniques whenever available.

Later, we will explain how packets belonging to different flows can become indistinguishable. Thus, we assume that as far as the eavesdropper is concerned, a packet transmission by a node may be for an original packet or just a relayed packet. Sensor nodes are usually vulnerable to capture and intrusion. An adversary is assumed to gain complete control of a compromised node including its routing information, collected data and all its cryptographic material. Therefore, it is important that no node is aware of the locations and identities of the sink or other data sources. In this paper we are only concerned with eavesdropping actions (passive attacks) of an adversary who is trying to locate a target node. Such an adversary usually prefers to be hidden, thus it refrains from executing active attacks e.g. denial of service, impersonation, jamming and so on, which may be discovered by the network operator using intrusion detection techniques.

5 Proposed Routing Scheme

In this section, we describe our Destination-Controlled Anonymous Routing Protocol for Sensors (DCARPS).

Part 1: Initialization
Before deploying a node i (including the sink), it is assigned a unique network identifier S_i. The sink and a sensor S_i are pre-programmed with a unique shared secret key K_i. During the lifetime of a sensor, its key may be updated by the sink. Due to computational and energy limitations of sensors, we only use symmetric cryptography[3]. Before deployment, the sink is also programmed to share a value with each sensor, denoted by DI_i (for Downstream Incoming), whose use in downlink communications will be explained later.

Part 2: Topology Discovery
Upon network activation and also periodically thereafter, a subset of sensors (or all of them in the worst case) broadcast route discovery messages. A sensor that originates a route discovery message includes its identity (ID) and a globally unique sequence number[4] in this message. All of the forwarding sensors append their IDs to this message. Each sensor repeats a route discovery message with a certain sequence number only once. All of these messages eventually, and usually in multiple copies, reach the sink allowing it to obtain a global view of the network topology. Please see Fig.2. A node that has already forwarded a route discovery message does not generate one of its own but will continue to relay such messages for other nodes.

This design is contrary to the common method of topology discovery in sensor networks based on broadcasting beacons from the sink. As explained before, that method violates location privacy of the sink. By making all the nodes, including the sink, behave in the same manner during topology discovery, we have effectively hid-

---

[3] Using light-weight cryptography, asymmetric cryptography on sensors is now possible as well.
[4] IETF RFC 4122 defines a namespace for globally unique identifiers.

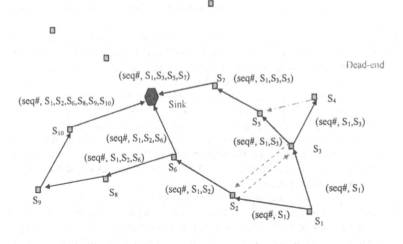

Fig. 2. Topology Discovery; S_1 broadcasts a route discovery message

den the sink. The details of our topology discovery protocol are published in another paper [13]. Currently, we are studying various techniques such as clustering and MultiPoint Relays in order to enhance the performance of this protocol.

Part 3: Route Calculation
Once the sink has acquired the network topology, it calculates routes for all the sensors. In the simplest case, it calculates only one route per sensor according to a pre-specified policy, such as "the shortest path". All the possible end-to-end routes for each sensor resemble multiple streams originating from the same point i.e. that sensor. Although each stream may branch out somewhere on the way but different branches of the eventual tree for each sensor can only merge at the final destination (the sink).

The final set of shortest paths is a tree structure rooted at the sink. Each sensor may have multiple downstream links but only one upstream link towards the sink. In other words, a *main branch* (a link connecting the sink to one of its neighbors) emanating from the sink may split several times into smaller branches, finally ending at individual leaves. A *leaf* is a sensor that does not have a downstream link on the shortest-paths tree. The shortest path for a leaf sensor contains the shortest paths for all its upstream sensors.

Part4: Uplink Path Establishment
In our routing protocol, the sink is responsible for establishing upstream next hops for all sensors. In this phase, the sink assigns labels for uplink communications to each node, including itself. A *label* is part of a packet that dictates how it should be forwarded. The label in an incoming packet at a node is called an incoming label, while the label in an outgoing packet is called an outgoing label. A node is the recipient of a packet if it has been assigned an incoming label that matches the label of the incom-

ing packet. To forward a packet, a node swaps the label in the packet with its own outgoing label. Further into the manuscript, we will explain how the sink communicates these label assignments to sensors.

Due to the tree-like structure of the routes, each sensor will have only one incoming label (even though it may have more than one incoming link) and one outgoing label in the upstream direction, because all sensors send their data to the same destination, namely the sink, and each one of them has only one upstream link. The outgoing label of a node is always the same as the incoming label of its upstream node. The sink may have no outgoing label, while the leaf sensors have no incoming labels. The sink can assign the same outgoing label to all of its neighbors. However, this may result in a large concentration of packets with the same label in its vicinity, which may be an indication to adversaries that the sink resides in that area. Therefore, we recommend that these labels must be different. A label assignment example for a simple shortest-paths tree is shown in Fig.3.

The sink uses *path_setup* messages to inform each sensor of its assigned uplink incoming label and outgoing label. These messages travel downwards along the branches from the sink to each sensor, according to the tree structure calculated previously by the sink. The format of these messages and how we maintain their secrecy is explained below

In order to reduce overhead, we distribute labels to all the sensors connected to one main branch by using just one message. The sink assembles a cryptographic onion per each main branch and broadcasts it locally. Each layer of an onion normally contains the incoming label and the outgoing label for the sensor that can successfully decrypt that layer. Each layer is encrypted by the sink using the secret key that it shares with the intended recipient. The layers are in such an order that one of the immediate neighbors of the sink can peel off the outermost layer of an onion. Each sensor broadcasts the rest of the onion after peeling off the outer layer. At each step, only one sensor can decrypt the outer layer. However, where a branch splits up, the corresponding sensor must broadcast as many sub-onions as the number of its down-

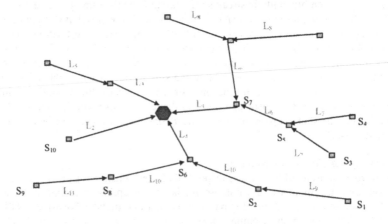

Fig. 3. Label assignments in a shortest-paths tree

stream sensors. These sub-onions are prepared by the sink in a way that the outer layer of each can only be decrypted by one of the sensors downstream from the branching sensor. As an example, the path_setup message broadcasted on the main branch labeled with L_4 in Fig.3 is:

$$K_6 (L_{10}, L_4, K_2 (L_9, L_{10}, K_1 (-, L_9)), K_8 (L_{11}, L_{10}, K_9(-, L_{11})))$$

$K_i (x)$ denotes encryption of x using the key K_i. At each layer, starting from the left, the first item is the incoming label and the next item is the outgoing label.

One possible algorithm for assembling the path_setup message for a main branch is the following: the sink starts at any leaf sensor and includes its outgoing label in the innermost layer of an onion. Then, it moves upstream and for each sensor adds a layer to the onion containing its incoming and outgoing labels. If it encounters a "branching sensor" (a sensor from where multiple branches start) it marks the onion so far built as a sub-onion. It then moves down on each of other branches, and for each one starts building a sub-onion beginning each at a leaf sensor. Once all the necessary sub-onions have been built, it puts them in a tandem and moves upstream from the branching sensor. This process continues until all the sensors in that main branch have been included in the onion. Please note that the sink applies this algorithm offline, to the routing information that it has collected. When all of the onions constructed in this way are broadcasted in the network, each node knows its incoming and outgoing labels.

Assignment of Unique Labels
As Fig.4 shows, when a node (B) broadcasts a packet, all nodes in its one-hop neighborhood receive that packet. Therefore, the nodes in the 2-hop neighborhood of the recipient (A) get the packet.

In order to avoid recipient ambiguity in a local broadcast, the incoming label of a node must be unique at least in its 2-hop neighborhood. In the label assignment phase, the sink must execute an algorithm to ensure that each node is allocated a unique incoming label. Below, we provide a heuristic algorithm for this task.

Denote the set of 2-hop neighbors of node X including itself with $N^2(X)$ and the yet unused labels in its 2-hop neighborhood with $L(X)$. At the start of the algorithm, the latter set is the same for all nodes and contains all the available labels. At every step

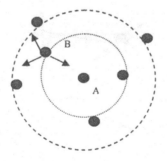

Fig. 4. A local broadcast

of the algorithm, the sink chooses a node (randomly or systematically) and assigns to it one of the labels still available in its label set. Then, it removes this label from the set of labels of that node and all its neighbors in a 2-hop area. One run of the algorithm ends when all the nodes have been allocated an incoming label. At the beginning of the next run, the set of unused labels for each node is re-initialized to contain all the available labels. In the following pseudocode, N is the set of all nodes and l_i is the incoming label assigned to node i.

$$
\begin{aligned}
&\textit{for all } i \in N \\
&\{ \\
&l_i = x; \quad //x \in L(i) \\
&\textit{for all } j \in N^2(i) \\
&\quad\quad L(j) = L(j) - \{x\}; \\
&\}
\end{aligned}
$$

Part 5: Uplink Data Transmission
When a sensor has a packet, for example some sensory data to report, it encrypts it for the sink. Then, it appends its outgoing label to the packet. The upstream neighbor, with an incoming label matching the packet label, accepts the packet and switches its label to its own outgoing label. At this point, it can rebroadcast the packet. However, to prevent a global eavesdropper that applies content analysis to the payload in order to trace the packet back to the source and/or destination, we make the payload to appear different at each hop. To do this, the forwarding sensor encrypts the already encrypted payload for the sink. Every sensor in the path repeats the encryption process, effectively creating an onion. Therefore, on each hop, the packet contains an unencrypted label and a payload that has been encrypted several times. Upon receiving the packet, the sink performs recursive decryptions to recover the original data.

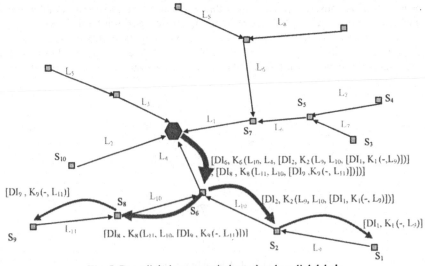

Fig. 5. Downlink data transmission using downlink labels

Part 6: Downlink Data Transmission
Downlink communications (from the sink to the sensors) is also based on layered cryptography and label switching. We explained before how the sink sends path_setup messages to all the sensors that belong to the same main branch in the shortest-paths tree. Other messages, such as acknowledgments and other control commands are sent in a similar manner. Remember that each node is initialized with a DI (Downstream Incoming) label. When the sink wishes to transmit a packet, it labels the outermost layer of an onion with the DI of one of its neighbors that is supposed to be the next hop for that packet according to the latest routing information available to the sink. However, inside that layer, which is encrypted for that neighbor, it precedes each sub-onion with the DI label of the next hop on the route.
As an example, the path setup process for the example in Fig.3 is illustrated in Fig. 5. For enhanced security, DI labels may be changed periodically.

Sink Anonymity in Downlink Communications
In order to prevent an eavesdropper from acquiring valuable information regarding the location of the sink, path_setup messages and other control packets sent by the sink in the downlink must be indistinguishable from data packets transmitted by sensors. Data packets consist of a label and an encrypted payload that is re-encrypted on every hop. Control packets look exactly like data packets since they have a label (DI) and an encrypted part that looks different on each hop because of layered cryptography. Moreover, these packets are only transmitted within a short setup phase, which gives the adversary little chance to find the sink while in other protocols any compromised node gives a clue about the sink's location.
Upon hearing a packet, a sensor performs the following algorithm:

> *If (label == my DI)*
>   *then*
>   *downlink packet, I am the immediate recipient.*
>   *decrypt payload with key shared with sink.*
>   *re-broadcast sub-onion(s).*
>   *else if (label exists in my routing table)*
>   *then*
>   *data packet, I am the immediate recipient.*
>   *swap label, (re)encrypt payload and forward packet.*
>   *else*
>   *discard packet.*

6...Summary

We have used the concepts of onion routing and label switching to develop an anonymous untraceable routing protocol for wireless sensor networks. Our protocol ensures location privacy for the sink by putting the control of routing in its hands. Also, unlike other routing protocols for sensor networks, in order to hide the location of the sink, our topology discovery process is initiated by the sensors not by the sink. To thwart content analysis attacks aimed at tracing packets back to the source or destination, we have used layered cryptography to make packets appear randomly different on different hops along their paths. However, in order to conserve the energy of sensors, we have used labels to identify the next hop for each packet so that only the immediate recipient of a packet attempts to decrypt it. Moreover, a sensor performs a minimal amount of computation and has only one key, which is shared with the sink.

References

1. Pandurang Kamat, Yanyong Zhang, Wade Trappe, Celal Ozturk, Enhancing Source-Location Privacy in Sensor Network Routing, in the proceedings of the 25th IEEE International Conference on Distributed Computing Systems (ICSCS'05), Vol. 00, pp. 599 – 608.
2. Xiaoyan Hong, Pu Wang, Jiejun Kong, Qunwei Zheng, Jun Liu, Effective Probabilistic Approach Protecting Sensor Traffic, IEEE Military Communication Conference (Milcom) 2005, Atlantic City, NJ, Oct 2005.
3. Shu Jiang, Nitin H.Vaidya, Wei Zhao, Routing in Packet Radio Networks to Prevent Traffic Analysis, in the proceedings of the IEEE Information Assurance and Security Workshop, West Point, NY, July 2000.
4. Jing Deng, Richard Han, Shivakant Mishra, Intrusion Tolerance and Anti-Traffic Analysis Strategies For Wireless Sensor Networks, in the proceedings of the IEEE International Conference on Dependable Systems and Networks (DSN 2004), 28 June - 1 July 2004, Florence, Italy.
5. Jing Deng, Richard Han, Shivakant Mishra, Countermeasures Against Traffic Analysis Attacks in Wireless Sensor Networks, First International Conference on Security and Privacy for Emerging Areas in Communications Networks (SECURECOMM'05), Athens, Greece, pp. 113-126.
6. Stephan Olariu, Mohamed Eltoweissy, Mohamed Younis, ANSWER: Autonomous Wireless Sensor Network, Q2SWinet'05, October 13, 2005, Montreal, Quebec, Canada.
7. Jamal N. Al-Karaki, Ahmed E. Kamal, Routing Techniques In Wireless Sensor Networks: A Survey, IEEE Wireless Communications, Vol. 11, No. 6. (2004), pp. 6-28.
8. J. Kong, Anonymous and Untraceable Communications in Mobile Wireless Networks, PhD thesis, University of California, Los Angeles, June 2004.
9. Y. Zhang, W. Liu, and W. Lou, Anonymous Communications in Mobile Ad Hoc Networks, IEEE INFOCOM, Miami, FL, March 2005.
10. J. Al-Muhtadi, R. Campbell, A. Kapadia, M. Dennis Mickunas, Seung Yi, Routing Through the Mist: Privacy Preserving Communication in Ubiquitous Computing Environments, in the proceedings of the 22nd International Conference on Distributed Computing Systems (ICDCS'02), July 2002, Vienna, Austria, p74.
11. A. Pfitzmann and M. Kohntopp. Anonymity, Unobservability and Pseudonymity -- A Proposal for Terminology, In Hannes Federath (Ed.), Designing Privacy Enhancing Technologies, Lecture Notes in Computer Science, LNCS 2009, pp. 1-9, Springer-Verlag, 2001.
12. D. L. Chaum, Untraceable electronic mail, return addresses, and digital pseudonyms, Communications of the ACM, 24(2), pp. 84–88, 1981.
13. Alireza A. Nezhad, Dimitris Makrakis, Ali Miri, Anonymous Topology Discovery for Wireless Sensor Networks, submitted to the 3-rd ACM International Workshop on QoS and Security for Wireless and Mobile Networks, October 22 - 26, 2007, Chania, Crete Island, Greece.

Model Checking Wireless Sensor Network Security Protocols: TinySec + LEAP *

Llanos Tobarra, Diego Cazorla, Fernando Cuartero, Gregorio Díaz, and Emilia Cambronero

Escuela Politécnica Superior de Albacete
Universidad de Castilla-La Mancha. 02071 Albacete, Spain.
{mtobarra, dcazorla, fernando, gregorio,emicp}@dsi.uclm.es

Abstract. In this paper, a formal analysis of security protocols in the field of wireless sensor networks is presented. Two complementary protocols, TinySec and LEAP, are modelled using the high-level formal language HLPSL, and verified using the model checking tool Avispa, where two main security properties are checked: authenticity and confidentiality of messages. As a result of this analysis, two attacks have been found: a man-in-the-middle- attack and a type flaw attack. In both cases confidentiality is compromised and an intruder may obtain confidential data from a node in the network. Two solutions to these attacks are proposed in the paper.

Keywords:Wireless sensor, model checking, security protocols, avispa toolbox

1 Introduction

Security has become a challenge in wireless sensor networks. Low capabilities of devices, in terms of computational power and energy consumption, make difficult using traditional security protocols.

Two main problems related to security protocols arise. Firstly, the overload that security protocols introduce in messages should be reduced at a minimum; every bit the sensor sends consumes energy and, consequently, reduces the life of the device. Secondly, low computational power implies that special cryptographic algorithms that require less powerful processors need to be used. The combination of both problems lead us to a situation where new approaches or solutions to security protocols need to be considered. These new approaches take into account basically two main goals: reduce the overhead that protocol imposes to messages, and provide reasonable protection while limiting use of resources.

* This work has been supported by the Spanish government with the project "Application of Formal Methods to Web Services", with reference TIN2006-15578-C02-02, and the JCCM regional project "Application of formal methods to the design and analysis of Web Services and e-commerce " (PAC06-0008-6995)

Please use the following format when citing this chapter:

Tobarra, L., Cazorla D., Cuartero, F., Díaz, G., Cambronero, E., 2007, in IFIP International Federation for Information Processing, Volume 248, Wireless Sensor and Actor Networks, eds. L. Orozco-Barbosa, Olivares, T., Casado, R., Bermudez, A., (Boston: Springer), pp. 95-106.

In order to design a secure network, several aspects have to be considered [1]: Key establishment and trust setup, secrecy and authentication, and privacy. Key establishment can be considered the base of the system; a secure and efficient key distribution mechanism is needed for large scale sensor networks. Once every node has its own keys, these are used to authenticate and encrypt (if needed) the messages they exchange. Several protocols have been proposed in the literature related to authentication and privacy [2,3], and key distribution [4,5,6]

In this paper we have focused in two of these protocols: TinySec [2] in the field of authentication and encryption, and LEAP [4] in the field of key management. TinySec is a fully-implemented link layer security architecture for wireless sensor networks included in the official TinyOS [7] release. LEAP is a powerful keying mechanism that supports the establishment of four types of keys for each sensor node with little intervention of the base station. A brief overview of these protocols is given in Section 2.

In line with the development of security protocols, some techniques have also been developed to model a system and check properties on it. One of the most promising techniques in this line is *model checking*. Model checking [8] is a formal methods based technique for verifying finite-state-concurrent systems, and has been implemented in several tools. One of the main advantages of this technique is that it is automatic and allows us to see if a system works as expected. In case the system does not work properly, the model checking tool provides a trace that leads to the source of the error.

Model checking has become a key point in the design of concurrent and distributed system because it allows us to ensure the correctness of a design at the earliest stage possible. Model checking has two main advantages over two classical techniques such as simulation and testing: *i)* we do not need to build a prototype of the system, and *ii)* we are able to verify the system against every single execution trace. The latter is very important because using simulation or testing we can only find errors, but we cannot ensure that the whole system behaves as expected (some errors may remain hidden until the system is in production stage).

Some general purpose model checking tools have been developed by different research groups: Spin, UPPAAL, Murϕ. These tools allow us to verify not only the functional properties of a system (e.g. Spin), but also the performance of a real-time system (e.g. UPPAAL). Although we can use these general purpose tools in order to verify security protocols, we consider that it is preferable (and more intuitive) to use a tool devoted to the verification of security protocols. Among these tools we can find Casper/FDR2 toolbox [9] and AVISPA [10].

The use of model checking tools to verify security protocols has been successful in the past in different areas such as Web Services [11,12,13], Instant Messaging [14], or Transport Layer security protocols [15,16].

In this paper, we present a formal verification of wireless sensor security protocols using AVISPA (Automated Validation of Internet Security Protocol and Applications) framework. AVISPA provides a high-level formal language HLPSL [17] for specifying protocols and their security properties. Once we have specified the model of the system, AVISPA translates it into an intermediate format IF. This is the input of several backends that are integrated into AVISPA framework: SATMC [18] (the one used in this analysis), OFMC, Cl-Atse and TA4SP. Besides, only one model is specified although it can be analysed with the four backends. AVISPA also offers a graphical interface SPAN [19] that helps in the specifying task.

Providing adequate security in wireless networks is not an easy task due to its broadcast nature. An intruder can overhear, intercept messages, inject new messages or modify messages in transit. This kind of intruder is called Dolev-Yao Intruder [20]. The intruder implemented in AVISPA is a Dolev-Yao intruder, which is appropriate to analyse wireless security protocols.

The paper is organised as follows. In Section 2 a brief overview of TinySec and LEAP is given. Section 3 is devoted to the formal verification of TinySec and LEAP, where four scenarios have been considered depending on the key distribution mechanism used. Finally in Section 4 we give our conclusions and future work.

2   TinySec + LEAP

TinySec [2] is a fully-implemented link layer security architecture for wireless sensor networks. The design of TinySec was based on existing security primitives proven to be secure. Using these primitives, a lightweight design was made taking into account wireless sensor networks particularities, mainly limited computation and communication capabilities, as well as low power consumption. TinySec is part of the official TinyOS [7] release.

The main goals of TinySec are performance, usability and security. Inside the security aspects, three main goals are considered: access control, message integrity, and confidentiality. Outside the scope of TinySec is to avoid replay attacks, which is left to higher layers of the protocol stack.

TinySec considers two operations with application layer data: authentication and semantically secure encryption. TinySec authenticates a packet using a message authentication code, CBC-MAC [21]. Semantically secure encryption[1] is made using an 8 byte initialisation vector (IV) and cipher block chaining (CBC) as encryption scheme [22]. Taking into account both operations, two kinds of packets can be found: TinySec-AE, that offers authentication and encryption, and TinySec-Auth, that offers only authentication. A detailed view of both kinds of packets is shown in Fig. 1 (see [2]).

In order to encrypt and decrypt data, shared keys are needed. TinySec does not address the problem of obtaining those keys; any particular keying mechanism can be used in conjunction with TinySec. In this paper we consider Localized Encryption and Authentication Protocol (LEAP) [4] as the keying mechanism. LEAP supports the establishment of four types of keys for each sensor node: an individual key shared with the base station, a pairwise key shared with

---

[1] Encrypting the same plaintext two times should give two different cipher-texts

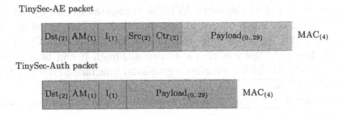

Fig. 1. TinySec packet formats: TinySec-AE and TinySec-Auth

another sensor node, a cluster key shared with multiple neighboring nodes, and a group key that is shared by all the nodes in the network. One interesting feature of LEAP is that it minimises the involvement of the base station. Fig. 2 shows the four LEAP keying mechanisms.

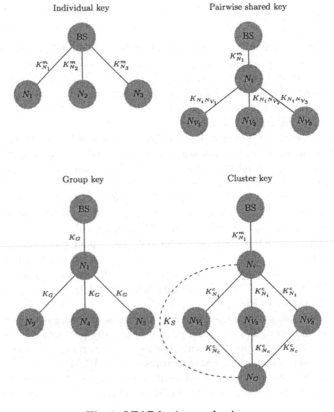

Fig. 2. LEAP keying mechanisms

3 Verification of TinySec + LEAP

The combination of TinySec and LEAP allows us to build a complete solution, where LEAP is responsible for obtaining the most convenient shared key at every moment, and TinySec is responsible for the authentication and encryption of messages exchanged between nodes. Thus, we have considered four different configurations of TinySec and LEAP, depending on the key mechanism used, and the kind and the situation of the nodes that communicate with each other:

- Case A. Request from the base node to a normal node using an individual key using messages TinySec-AE.
- Case B. Request from the base node to a normal node using an individual key using messages TinySec-Auth.
- Case C. Communication between immediate neighbouring nodes using a pairwise shared key.
- Case D. Communication between immediate neighbouring nodes using a cluster key.
- Case E. Communication between non immediate neighbouring nodes nodes using a cluster key.

During all these analysis we will consider that before deployment, a node master key K_m has been saved inside every node in the network. We also adopt the Dolev-Yao intruder model, where an intruder can overhear, intercept, alter, or inject any messages into the radio communication channel.

Case A. The network configuration is shown in Fig. 3. In this case a base node (BS) makes a request to a normal node (N_i) which has an individual (unique) key that it shares with the base station

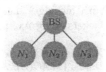

Fig. 3. Sensor network: cases A and B.

The protocol for TinySec-AE packets using Avispa syntax is as follows:

1. BS $\rightarrow$ N_i: $IV_1\{(IV_1 \oplus Data_1)\}_{F(K_m.N_i)}$. $\{MAC(IV_1.Data_1)\}_{H(K_m.N_i)}$
   where $IV_1 = N_i.AM_1.Size_1.BS.Counter$
2. N_i $\rightarrow$ BS: $IV_2.\{(IV_2 \oplus Data_2)\}_{F(K_m.N_i)}$. $\{MAC(IV_2.Data_2)\}_{H(K_m.N_i)}$
   where $IV_2 = BS.AM_2.Size_2.N_i.Suc(Counter)$

In this protocol, functions F and H are pseudo-random functions that allow to calculate the encryption key and the authentication key taking as parameters the master key (K_m) and the id of the node (N_i). Due to the fact that Avispa does not offer arithmetic semantics, the increase of the counter is represented by a function Suc, such as $Suc(0)$ represents 1, $Suc(Suc(0))$ represents 2 and so on.

The properties we have to analyse are the following:

- Authentication of $Data_1$ and $Data_2$. i.e., the node N_i and the base station (BS) share the same value for $Data_1$ and $Data_2$ and both execute the same session of the protocol. This property allows us to proof that bilateral authentication is achieved by using the MAC, and the integrity of the message is guarantied.
- Confidentiality of $Data_1$ and $Data_2$, i.e., $Data_1$ and $Data_2$ are secret values shared between N_i and BS, and they are not known by an intruder or third parties.

The verification with Avispa finds only the following replay attack, where I_{BS} represents an intruder playing the role of the base station:

1. BS $\rightarrow$ N$_1$: $IV_1.\{(IV_1 \oplus Data_1)\}_{F(K_m.N_1)}. \{MAC(IV_1.Data_1)\}_{H(K_m.N_1)}$
   where $IV_1 = N_1.AM_1.Size_1.BS.0$
2. N$_1$ $\rightarrow$ BS : $IV_2.\{(IV_2 \oplus Data_2)\}_{F(K_m.N_1)}. \{MAC(IV_2.Data_2)\}_{H(K_m.N_1)}$
   where $IV_2 = BS.AM_2.Size_2.N_1.Suc(0)$
1. I_{BS} $\rightarrow$ N$_1$: $IV_1.\{(IV_1 \oplus Data_1)\}_{F(K_m.N_1)}. \{MAC(IV_1.Data_1)\}_{H(K_m.N_1)}$
   where $IV_1 = N_1.AM_1.Size_1.BS.0$
2. N$_1$ $\rightarrow$ BS : $IV_3.\{(IV_3 \oplus Data_3)\}_{F(K_m.N_1)}. \{MAC(IV_3.Data_3)\}_{H(K_m.N_1)}$
   where $IV_3 = BS.AM_3.Size_3.N_1.Suc(0)$

Nevertheless, as was said before, TinySec does not manage replay attacks, which are left to higher layers of the protocol stack. Apart from this attack, the protocol is secure, even when a node is compromised by the intruder.

Case B. In this case we use the same scenario than in the previous case (see Fig. 3), but we consider TinySec-Auth messages instead of TinySec-AE messages. The protocol for TinySec-AE packets using Avispa syntax is as follows:

1. BS $\rightarrow$ N$_i$: $N_i.AM_1.Size_1.Data_1. \{MAC(N_i.AM_1.Size_1.Data_1)\}_{H(K_m.N_i)}$
2. N$_i$ $\rightarrow$ BS: $BS.AM_2.Size_2.Data_2. \{MAC(BS.AM_2.Size_2.Data_2)\}_{H(K_m.N_i)}$

As we mentioned before, TinySec-Auth does not provide any confidentiality mechanisms. Thus, we can only analyse the authentication of $Data_1$ and $Data_2$, i.e., we can proof the bilateral authentication between BS and N_i, by means of the MAC messages, and also the integrity of messages. As in the previous case, we have found a replay attack that we omit

Case C. The network configuration is shown in Fig. 4. In this case a node (N_1) shares a pairwise key with each of its immediate neighbours (N_2 and N_3). The protocol for TinySec-AE packets using Avispa syntax is as follows, where A represents node N_1 and B represents one of its neighbours (N_2 or N_3):

1. $A \rightarrow B : A, Nonce_A$
2. $B \rightarrow A : B.\{MAC(Nonce_A.B)\}_{H(K_m.B)}$
3. $A \rightarrow B : B.AM_1.Size_1.Data_A.\{MAC(B.AM_1.Size_1.Data_A)\}_{K'_{AB}}$
   where $K'_{AB} = H(H(Km.B).A)$
4. $B \rightarrow A : IV_2.\{(IV_2 \oplus Data_B)\}_{K_{AB}}. \{MAC(IV_2.Data_B)\}_{K'_{AB}}$
   where $IV_2 = A.AM_2.Size_2.B.Counter$
   and $K_{AB} = F(F(Km.N_2).N_1)$

Fig. 4. Sensor network: cases C and D

The properties we have to analyse are the following:

- Authentication of $Nonce_A$, $Data_A$ and $Data_B$. i.e., nodes A and B share the same value for $Nonce_A$, $Data_A$ and $Data_B$ and both execute the same session of the protocol.
- Confidentiality of $Data_B$, i.e., $Data_B$ is a secret value shared between A and B, and remains unknown to an intruder or third parties.

In this case Avispa finds a *man-in-the-middle* attack, where I_A represents the intruder playing the role of node A, I_{B_2} represents the intruder playing the role of node B, B_1 represents node B communicating with the intruder, and B_2 represents node B communicating with A:

1. $I_A \rightarrow B_1 : A, Nonce_I$
2. $B_1 \rightarrow I_A : B.\{MAC(Nonce_I.B)\}_{H(K_m.B)}$
1. $A \rightarrow I_{B_2} : A, Nonce_A$
2. $I_A \rightarrow B_2 : N_x.Nonce_A$
2. $B_2 \rightarrow A : B.\{MAC(Nonce_A.B)\}_{H(K_m.B)}$
3. $A \rightarrow I_{B_2} : B.AM_1.Size_1.Data_A.\{MAC(B.AM_1.Size_1.Data_A)\}_{K'_{AB}}$
3. $I_A \rightarrow B_1 : B.AM_1.Size_1.Data_A.\{MAC(B.AM_1.Size_1.Data_A)\}_{K'_{AB}}$
4. $B_1 \rightarrow I_A : A.AM_2.Size_2.B.Counter.\{(IV_2 \oplus Data_B)\}_{K_{AB}}.$
   $\{MAC(IV_2.Data_B)\}_{K'_{AB}}$

First the intruder, playing the role of A, starts the protocol with B (denoted B_1), and sends a false nonce, which is answered by B (B_1). Then, A starts a session with B (B_2) but this message is intercepted by the intruder (I_{B_2}) which modifies the message and sends to B (B_2) the identity of a false node N_x and the nonce of A. Node B (B_2) sends the answer of the last message to A, and A responds with a request of data ($Data_2$) to B (B_2). Again, the request is intercepted by the intruder who redirects the message to B (B_1). At this moment, B (B_1) thinks that it has received a correct request from A, and then it sends $Data_B$ to the intruder playing the role of A (I_A)

At the end we conclude that B_1 has exchanged information ($Data_2$) with the intruder, and B_2 thinks that it has talked to a node N_x that does not exist. A solution to this attack consists in authenticate no only the answer from B in message 2 but also the message 1 sent from A. The modified version of the protocol is:

1. $A \rightarrow B : A.Nonce_A.MAC(A.Nonce_A)_{H(K_m,A)}$
2. $B \rightarrow A : B.\{MAC(Nonce_A.B)\}_{H(K_m.B)}$
3. $A \rightarrow B : B.AM_1.Size_1.Data_A.\{MAC(B.AM_1.Size_1.Data_A)\}_{K'_{AB}}$
   where $K'_{AB} = H(H(Km.B).A)$
4. $B \rightarrow A : IV_2.\{(IV_2 \oplus Data_B)\}_{K_{AB}}.\{MAC(IV_2.Data_B)\}_{K'_{AB}}$
   where $IV_2 = A.AM_2.Size_2.B.Counter$ y $K_{AB} = F(F(Km.N_2).N_1)$

Case D. Sensor network configuration is shown in Fig. 4. In this case a node (N_1) shares a cluster key with each of its immediate neighbours (N_2 and N_3).

The protocol for TinySec-AE packets using Avispa syntax is as follows, where A represents node N_1, B represents one of its neighbours (N_2 or N_3), and K_c is the cluster key:

1. $A \rightarrow B : B.AM_1.Size_1.A.Counter.\{(IV_1 \oplus K_c)\}_{K_{AB}}.\{MAC(IV_1.K_c)\}_{K'_{AB}}$
   where $IV_1 = B.AM_1.Size_1.A.Counter$, $K_{AB} = F(F(Km.B).A)$ and
   $K'_{AB} = H(H(Km.B).A)$
2. $B \rightarrow A : BS.AM_2.Size_2.done.\{MAC(BS.AM_2.Size_2.done)\}_{H(K_c)}$
3. $A \rightarrow B : B.AM_1.Size_1.Data_A.\{MAC(B.AM_1.Size_1.Data_A)\}_{H(K_c)}$
4. $B \rightarrow A : IV_2.\{(IV_2 \oplus Data_B)\}_{F(K_c)}.\{MAC(IV_2.Data_B)\}_{H(K_c)}$
   where $IV_2 = A.AM_2.Size_2.B.Suc(Counter)$

The properties we have to analyse are the following:

- Authentication of K_c, $Data_A$ and $Data_B$. i.e., nodes A and B share the same value for K_c, $Data_A$ and $Data_B$ and both execute the same session of the protocol.
- Confidentiality of $Data_B$ and K_c, i.e., $Data_B$ and K_c are secret values shared between A and B, and they remain unknown to an intruder or third parties.

After analysing the protocol with Avispa, an interesting attack based on types was found:

1. $A \rightarrow B : B.AM_1.Size_1.A.Counter.\{(IV_1 \oplus K_c)\}_{K_{AB}}.\{MAC(IV_1.K_c)\}_{K'_{AB}}$
   where $IV_1 = B.AM_1.Size_1.A.Counter$,
   $K_{AB} = F(F(Km.B).A)$ and $K'_{AB} = H(H(Km.B).A)$
2. $B \rightarrow I_A : BS.AM_2.Size_2.done.\{MAC(BS.AM_2.Size_2.done)\}_{H(K_c)}$
3. $I_A \rightarrow B : BS.AM_2.Size_2.done.\{MAC(BS.AM_2.Size_2.done)\}_{H(K_c)}$
4. $B \rightarrow I_A : IV_2.\{(IV_2 \oplus Data_B)\}_{F(K_c)}.\{MAC(IV_2.Data_B)\}_{H(K_c)}$
   where $IV_2 = A.AM_2.Size_2.B.Suc(Counter)$

In this attack the intruder intercepts the message sent from B to A in step 2. In step 3, the intruder sends the intercepted message back to B as if was a true request from A to B. B takes the message as a request an misunderstands the label done as it was $Data_A$. In step 4, B sends $Data_B$ to the intruder.

This is a type flaw attack, i.e., type checking has not been done and a constant label has been interpreted as a variable data. One solution to this attack has been proposed by Heather et al. in [23] which basically consists in tagging each field with information about its type, although this solution could not be adequate in wireless sensor networks because it adds some extra bits of information into each message. In real implementations of the protocol, programmers should take into account this type flaw attack and do type checking in order to avoid a possible attack. In any case, this kind of attack can be a problem because network bandwidth could be saturated and there is a consumption of resources in the node.

Case E. Sensor network configuration is shown in Fig. 5. In this case a node (e.g. N_1) shares a cluster key with non immediate neighbouring nodes (e.g. N_4 and N_5).

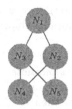

Fig. 5. Sensor Network: Case E

The protocol for TinySec-AE packets using Avispa syntax is as follows, where A represents node N_1, B represents a non neighbour node (N_4 or N_5), N_i represents a neighbour node (N_2 and N_3):

Q. $A \rightarrow N_i : A, B$

R. $N_i \rightarrow A : A.AM_1.Size_1.N_i.\{MAC(A.AM_1.Size_1.N_i)\}_{K'_{AN_i}}$

1. $A \rightarrow N_i : IV_2.F(SK_i,0).\{(IV_2 \oplus SK_i)\}_{K_{AN_i}} \cdot \{MAC(IV_2.SK_i.F(SK_i,0))\}_{K'_{AN_i}}$
   where $IV_2 = N_i.AM_2.Size_2.A.Counter$

2. $N_i \rightarrow B : IV_3.F(SK_i,0).\{(IV_3 \oplus SK_i)\}_{K_{N_i B}} \cdot \{MAC(IV_3.SK_i.F(SK_i,0))\}_{K'_{N_i B}}$
   where $IV_3 = B.AM_3.Size_3.N_i.Counter$
   When B has every SK_i

D. $B \rightarrow A : A.AM_4.Size_4.done\{MAC(A.AM_4.Size_4.done)\}_{Sk}$
   where $Sk = Sk_1 \oplus \ldots \oplus Sk_n$

Avispa only finds a replay attack, that we do not consider. Apart from this attack, the protocol is secure.

4 Conclusions and Future Work

In this paper we have presented a formal approach to the security analysis of wireless sensor networks by means of a model checking tool called Avispa. Several models of the network have been considered depending on the relative position and roles of the nodes. Two wireless sensor security protocols have been considered in order to build a complete solution: TinySec, who is in charge of the authentication and encryption of messages, and LEAP, who covers the key distribution mechanism.

The five models we have presented have been analysed with Avispa, and we have obtained the following results:

Case A. Request from the base node to a normal node using an individual key using messages TinySec-AE. The verification with Avispa finds only a replay attack where an intruder may play the role of the base station. Nevertheless, TinySec does not manage replay attacks, which are left to higher layers of the protocol stack. Apart from this attack, the protocol is secure,

Case B. Request from the base node to a normal node using an individual key using messages TinySec-Auth. As in the previous case, Avispa finds only a replay attack. Under our previous assumptions about replay attacks, we can consider that the protocol is secure,

Case C. Communication between immediate neighbour nodes using a pairwise shared key. In this case Avispa finds a *man-in-the-middle* attack where the intruder may play at the same time the role of two nodes in order to obtain real information from one of them. Consequently, confidentiality is lost. A solution to this attack is proposed and consists in authenticate the first message sent from the initiator (node *A*).

Case D. Communication between immediate neighbour nodes using a cluster key. In this case a type flaw attack is found. As in the previous case, the intruder can obtain real data from one of the nodes, and therefore confidentiality is lost. A solution to this kind of attack has been proposed in [23]; nevertheless this solution is not adequate in wireless sensor networks because increases the overload of security protocols in messages.

Case E. Communication between non immediate neighbour nodes using a cluster key. In this case, apart from a replay attack, the protocol is secure.

Our future work is concerned with extending our analysis to other security protocols for wireless sensor networks such as SNEP, μTESLA (both defined in [3]), and MiniSec [24]. Also we are interested in the analysis of TinySec with other distribution key protocols such as LEAP+[25] and TinyPK [26].

References

1. Perrig, A., Stankovic, J.A., Wagner, D.: Security in wireless sensor networks. Commun. ACM **47** (2004) 53–57

2. Karlof, C., Sastry, N., Wagner, D.: TinySec: a link layer security architecture for wireless sensor networks. In: Proceedings of the 2nd International Conference on Embedded Networked Sensor Systems, SenSys 2004, Baltimore, MD, USA, November 3-5, 2004, ACM (2004) 162–175

3. Perrig, A., Szewczyk, R., Tygar, J.D., Wen, V., Culler, D.E.: SPINS: Security protocols for sensor networks. Wireless Networks 8 (2002) 521–534

4. Zhu, S., Setia, S., Jajodia, S.: LEAP: efficient security mechanisms for large-scale distributed sensor networks. In Jajodia, S., Atluri, V., Jaeger, T., eds.: ACM Conference on Computer and Communications Security, ACM (2003) 62–72

5. Chan, H., Perrig, A., Song, D.X.: Random key predistribution schemes for sensor networks. In: IEEE Symposium on Security and Privacy, IEEE Computer Society (2003) 197

6. Eschenauer, L., Gligor, V.: A key-management scheme for distributed sensor networks. In Atluri, V., ed.: ACM Conference on Computer and Communications Security, ACM (2002) 41–47

7. Hill, J., Szewczyk, R., Woo, A., Hollar, S., Culler, D.E., Pister, K.S.J.: System architecture directions for networked sensors. In: Inter. Conf. on Architectural Support for Programming Languages and Operating Systems, ASPLOS. (2000) 93–104

8. Clarke, E.M., Grumberg, O., Peled, D.A.: Model Checking. The MIT Press (1999)

9. Lowe, G.: Casper: A compiler for the analysis of security protocols. Journal of Computer Security 6 (1998) 53–84

10. Armando, A., Basin, D.A., Boichut, Y., Chevalier, Y., Compagna, L., Cuéllar, J., Drielsma, P.H., Héam, P.C., Kouchnarenko, O., Mantovani, J., Mödersheim, S., von Oheimb, D., Rusinowitch, M., Santiago, J., Turuani, M., Viganò, L., Vigneron, L.: The AVISPA tool for the automated validation of internet security protocols and applications. In Etessami, K., Rajamani, S.K., eds.: CAV. Volume 3576 of Lecture Notes in Computer Science., Springer (2005) 281–285

11. Tobarra, M.L., Cazorla, D., Cuartero, F., Diaz, G.: Application of formal methods to the analysis of web services security. In Bravetti, M., Kloul, L., Zavattaro, G., eds.: EPEW/WS-FM. Volume 3670 of Lecture Notes in Computer Science., Springer (2005) 215–229

12. Backes, M., Mödersheim, S., Pfitzmann, B., Viganò, L.: Symbolic and cryptographic analysis of the secure WS-ReliableMessaging scenario. In Aceto, L., Ingólfsdóttir, A., eds.: FoSSaCS. Volume 3921 of Lecture Notes in Computer Science., Springer (2006) 428–445

13. Bhargavan, K., Fournet, C., Gordon, A.D.: Verifying policy-based security for web services. In Atluri, V., Pfitzmann, B., McDaniel, P.D., eds.: ACM Conference on Computer and Communications Security, ACM (2004) 268–277

14. Mannan, M., van Oorschot, P.C.: A protocol for secure public instant messaging. In Crescenzo, G.D., Rubin, A., eds.: Financial Cryptography. Volume 4107 of Lecture Notes in Computer Science., Springer (2006) 20–35

15. Mitchell, J.C.: Finite-state analysis of security protocols. In Hu, A.J., Vardi, M.Y., eds.: CAV. Volume 1427 of Lecture Notes in Computer Science., Springer (1998) 71–76

16. Tobarra, M.L., Cazorla, D., Cuartero, F., Diaz, G.: Formal verification of TLS handshake and extensions for wireless networks. In: Proc. of IADIS International Conference on Applied Computing (AC'06), San Sebastian, Spain, IADIS Press (2006) 57–64

17. Chevalier, Y., Compagna, L., Cuéllar, J., Drielsma, P.H., Mantovani, J., Möder-sheim, S., Vigneron, L.: A high level protocol specification language for industrial security-sensitive protocols. In: Proceedings of Workshop on Specification and Automated Processing of Security Requirements (SAPS). (2004) 193–205
18. Armando, A., Compagna, L.: SATMC: A SAT-based model checker for security protocols. In Alferes, J.J., Leite, J.A., eds.: JELIA. Volume 3229 of Lecture Notes in Computer Science., Springer (2004) 730–733
19. Glouche, Y., Genet, T., Heen, O., Courtay, O.: A security protocol animator tool for AVISPA. In: ARTIST2 Workshop on Security Specification and Verification of Embedded Systems, Pisa (2006)
20. Dolev, D., Yao, A.C.C.: On the security of public key protocols. In: FOCS, IEEE (1981) 350–357
21. Bellare, M., Kilian, J., Rogaway, P.: The security of the cipher block chaining message authentication code. J. Comput. Syst. Sci. 61 (2000) 362–399
22. Bellare, M., Desai, A., Jokipii, E., Rogaway, P.: A concrete security treatment of symmetric encryption. In: Proceedings of 38th Annual Symposium on Foundations of Computer Science, IEEE (1997) 394–403
23. Heather, J., Lowe, G., Schneider, S.: How to prevent type flaw attacks on security protocols. Journal of Computer Security 11 (2003) 217–244
24. Mark Luk, Ghita Mezzour, A.P., Gligor, V.: Minisec: A secure sensor network communication architecture. In: Proceedings of IEEE International Conference on Information Processing in Sensor Networks (IPSN). (2007)
25. Zhu, S., Setia, S., Jajodia, S.: LEAP : Efficient security mechanisms for large-scale distributed sensor networks. ACM Transactions on Sensor Networks 2 (2006) 500–528
26. Watro, R.J., Kong, D., Cuti, S.f., Gardiner, C., Lynn, C., Kruus, P.: TinyPK: securing sensor networks with public key technology. In Setia, S., Swarup, V., eds.: SASN, ACM (2004) 59–64

A Synchronous Engine for Wireless Sensor Networks

Fernando Royo, Teresa Olivares and Luis Orozco-Barbosa
Albacete Research Institute of Informatics
University of Castilla La Mancha
02071 Albacete, SPAIN
[froyo, teresa, lorozco]@dsi.uclm.es

Abstract. The development and deployment of robust and highly populated Wireless Sensor Networks require addressing a wide variety of challenges. Among them, the synchronization and energy management of the nodes composing the network have been identified as two challenges to overcome. In this paper, we follow a cross-layer approach by defining a synchronous engine built across the radio, MAC and routing layers. The design of our proposal has been based on the results of our previous research efforts dealing with various experimental platforms and experimental trials.

1 Introduction

Traditionally, an energy efficient wireless MAC protocol is a protocol that minimizes idle listening and overhearing [1]. In addition, as any other MAC protocol, it should keep to a minimum the number of collisions and the protocol overhead. Idle listening is a dominant factor on energy waste in most sensor network applications. The central approach to reducing energy lost to idle listening is to lower the radio duty cycle by turning the radio off part of the time. Duty cycle is the ratio between listen time and a full listen/sleep interval.

Radio duty cycling, where the radio is off by default but wakes up periodically to participate in potential network communication, has received significant attention in the literature. However, the duty-cycling benefits achieved in theory and simulations have often not translated to practice. This can be attributed mainly to the problem of time uncertainty between sensor nodes. If the sleep/wakeup schedules of nodes do not intersect, the communication can not take place. Note that each sensor node may have its own notion of time governed by its local clock. The approaches used by MAC protocols to address this problem of *time uncertainty* determine their energy consumption [8]. The lack of techniques to accurately estimate time uncertainty also impacts the ability to deploy long-lived sensor network applications.

Recent studies on MAC protocols for sensor networks observe that there is not clear trend indicating that medium access for sensor networks is converging towards a unique best solution [13]. Besides, many of the protocols being introduced in the literature have only been evaluated in simulation. In order to impact the market, a

Please use the following format when citing this chapter:

Royo, F., Olivares, T., Orozco-Barbosa, L., 2007, in IFIP International Federation for Information Processing, Volume 248, Wireless Sensor and Actor Networks, eds. L. Orozco-Barbosa, Olivares, T., Casado, R., Bermudez, A., (Boston: Springer), pp. 107-118.

protocol not only must perform well in simulation; it must also integrate well within the implementations of wireless sensor network protocol architectures

This paper presents SA-MAC (Synchronous after Awake MAC), a new MAC protocol for wireless sensor networks. Our objective is to address some of the major issues currently present in this type of networks. We are particular interested on those issues related to the lower layers, such as energy management, overhearing, packet overhead or idle listening. Furthermore, the synchronization utilized in the protocol denotes a cross layer view, fusing some characteristics from the network layer over the MAC layer.

This article presents this new and robust protocol, as a component of a complete network architecture and coexisting with other important protocols. The remainder of this article is structured as follows. Section 2 reviews the state-of-the art on MAC protocols for sensor networks. Section 3 shows our previous work with sensor networks, implementations and results. Section 4 describes our proposal. Section 5 concludes the paper and outlines our future work.

2 Background

Medium access for sensor networks is a very active research area. The B-MAC protocol has been an important input to the MAC protocols world for wireless sensor network. We show in the following table the most important protocols for wireless sensor networks in the last five years, and then we briefly explain the most important differences between them. Existing works mainly focus on two directions: TDMA and contention-based protocols. Two classes of contention-based protocols are those that add schedules and those that employ channel sampling. All protocols control the radio duty cycle to avoid energy waste in idle listening; some also take approaches to avoid overhearing or add other optimizations [10]. The major advantage of scheduling is that a sender knows a receiver's wakeup time and thus transmits efficiently. However, the cost of listening for an entire contention interval is about ten times the cost of channel sampling, thus the overhead in lightly used networks is higher then Low Power Listening (LPL) based approaches [4].

S-MAC, Sensor Media Access Control [1, 2], is a contention based MAC protocol that adds into the MAC layer power management, link-level retransmission, duplicate packet suppression, hidden terminal avoidance using RTS/CTS, and link-quality estimation. S-MAC periodically sleeps, wakes up, listens to the channel, and then returns to sleep. S-MAC is designed to operate like a "black box"—it is optimized for a representative set of workloads. S-MAC does not have any hooks to change its duty cycle or reconfigure its parameters; instead, it combines link, routing, organization, synchronization, and fragmentation services into a single protocol. All of these services may be used by network protocols. In the power management scheme in S-MAC, each active period is of fixed size, 115 ms, with a variable sleep period. The length of the sleep period dictates the duty cycle of S-MAC. At the beginning of each active period, nodes within a cell exchange synchronization information. S-MAC uses Adaptive Listen, which allows S-MAC to achieve lower latency when relaying multihop traffic. By changing the duty cycle, S-MAC can trade energy for latency [3].

Table 1. Mac Protocols

NAME	YEAR	UNIVERSITY	CHARACTERISTICS
S-MAC	2002	USC, USA	Scheduled Fixed duty cycle
T-MAC	2003	Delft, Holland	Scheduled Adaptive duty cycle
WISE-MAC	2004	CSEM, Switzerland	Synchronized preamble sampling
B-MAC	2004	UCB, USA	Adaptive preamble sampling (CCA & LPL)
Z-MAC	2005	NCSU, USA	hybrid protocol (TDMA/CSMA)
UB-MAC	2005	UM, USA	Uncertainty-driven time synchronization
B-MAC+	2006	Pisa, Italy	B-MAC wake-up preamble division
SCP-MAC	2006	USC, USA	LPL + Scheduling
Crankshaft	2007	Delft, Holland	SCP-MAC for Dense WSN
SA-MAC	2007	UCLM, Spain	Synchronization based packet and take advantage it for routing

To handle load variations in time and location T-MAC, Timeout MAC [5], introduces an adaptive duty cycle in a novel way: by dynamically ending the active part of it. This reduces the amount of energy wasted on idle listening, in which nodes wait for potentially incoming messages, while still maintaining a reasonable throughput. T-MAC in variable workload uses one fifth the power of S-MAC. In homogeneous workloads, T-MAC and S-MAC perform equally well. T-MAC suffers from the same complexity and scaling problems of S-MAC [3].

WiseMAC (Wireless Sensor MAC) [5] is based on the preamble sampling technique. This technique consists in regularly sampling the medium to check for activity. By sampling the medium, we mean listening to the radio channel for a short duration, e.g. the duration of a modulation symbol. All sensor nodes in a network sample the medium with the same constant period. Their relative sampling schedule offsets are independent and constant. This technique provides a very low power consumption when the channel is idle. The disadvantages of this technique are that the (long) wake-up preambles cause a throughput limitation and large power consumption overhead in reception. The novel idea introduced by WiseMAC consists in letting the access point learn the sampling schedule of all sensor nodes. Knowing the sampling schedule of the destination, the access point starts the transmission just at the right time with a wake-up preamble of minimized duration. WiseMAC differs from previous research on ad-hoc sensor networks, mainly because it focus on an infrastructure topology, and investigate how the unconstrained energy supply of the access point can be exploited.

B-MAC, Berkeley Media Access Control [4], a carrier sense media access protocol for wireless sensor networks that provides a flexible interface to obtain ultra low power operation, effective collision avoidance, and high channel utilization, is motivated by the needs of monitoring applications. B-MAC supports on-the-fly reconfigu-

ration and provides bidirectional interfaces for system services to optimize performance, whether it be for throughput, latency, or power conservation [3]. To achieve
low power goals and workload requirements, S-MAC is not only a link protocol, but
also a network and configuration protocol. Applications and services must rely on
policies internal to S-MAC to adjust operation as node and network conditions
change; such changes are opaque to the application. In contrast, the B-MAC protocol
contains a small core of media access functionality. B-MAC uses clear channel as-
sessment (CCA) and packet back offs for channel arbitration, link layer acknowledgments for reliability and low power listening (LPL) for low power communication.
To achieve low power operation, B-MAC employs an adaptive preamble sampling
scheme to reduce duty cycle and minimize idle listening. WiseMAC meets many of
B-MAC goals except that it has no mechanism to reconfigure based on changing
demands from services using the protocol [4].

Z-MAC [6] is a hybrid protocol for wireless sensor network that combines the
strengths of TDMA and CSMA while offsetting their weaknesses. Like CSMA, Z-
MAC achieves high channel utilization and low-latency under low contention and
like TDMA, achieves high channel utilization under high contention and reduces
collision among two-hop neighbours at a low cost. Z-MAC has the setup phase in
which it runs the following operations in sequence: neighbour discovery, slot assignment, local frame exchange and global time synchronization. These operations run
only once during the setup phase and does not run until a significant change in the
network topology occurs. Z-MAC uses the CCA and LPL features of B-MAC. Thus,
its energy efficiency is no better than B-MAC's under low data applications.

UB-MAC, uncertainty B-MAC [7], integrate an uncertainty-driven time synchronization scheme with B-MAC, and empirically demonstrate one o two orders of magnitude reduction in the transmit energy consumption at a node with negligible impact
on the packet loss rate. B-MAC uses an asynchronous technique that involves no time
synchronization or clock estimation to deal with the time uncertainty. Other techniques such as S-MAC and T-MAC use synchronized techniques where explicit time
synchronization beacons are transmitted periodically between neighbouring nodes.
This enables the transmitter to turn on the radio at the right moment, but the inability
to deal effectively with time varying changes in clock drift force these techniques to
re-synchronize frequently. In [7] the authors, first, experimentally obtained long time-
scale data sets both in indoor and outdoor settings for Berkeley mica2 motes. They
then performed a detailed characterization period on accuracy and energy requirements. Second, they used the results of the empirical analysis to design an adaptive
time synchronization protocol and, lastly, they developed a prototype implementation
on mica2 motes for sense-response applications (only with three nodes).

B-MAC+ [9] is an enhancement of B-MAC. The basic idea consists in replacing
the pattern of the wake-up preamble of B-MAC, with a new pattern that contains
information about the size of the remaining part of the preamble not yet transmitted.
This information can be used by receivers to avoid wait states during significant por-
tions of the preamble transmission time, i.e. going to sleep and waking up when the
data payload is actually transmitted. The wake-up preamble of B-MAC+ is obtained
by dividing into slots the wake-up preamble of B-MAC. The experiments were made
only with two Tmote nodes, one sender node and one receiver node connected to the

USB port of a PC running a modified version of TOS_Base. The results show statistics about the time the radio must be on to receive a packet, and relates the collected data with the analytical mean values of both B-MAC+ and B-MAC. B-MAC+ outperforms B-MAC, as the countdown packets technique reduces the waste of time in receiving long wake-up preambles.

SCP-MAC, Scheduled Channel Polling [10], eliminates long preambles in LPL for all transmissions, and is able to operate ultra-low duty cycles when traffic is light by synchronizing the channel polling times. It has been designed with two main goals: first, to push the duty cycle an order of magnitude lower than in practical with current MAC protocols and second, to adapt to variable traffic loads common in many sensor networks applications. To schedule coordinated transmission and listen periods is the approach of S-MAC and T-MAC. The schedule determines when a node should listen and when it should sleep. In S-MAC and T-MAC nodes adopt common schedules, synchronizing with periodic control messages. A receiver only listen to brief contention nodes participating in data transfer remain awake after contention periods, while others can then sleep. Overhead is due to schedule maintenance and listening during contention intervals if there is nothing to send. Another technique is LPL presented in WiseMAC and B-MAC. In LPL, nodes wake up very briefly to check channel activity without actually receiving data. The authors of SCP-MAC call this action channel polling (polling refers only to each node sampling the channel to check for activity). Unfortunately, current LPL-based protocols have some problems: the duty cycle is limited to 1-2% because the polling frequency needs to balance the cost on sending preambles and polling the channel; this balance between sender and receiver costs makes LPL-based protocols very sensitive to tuning for an expected neighbourhood size and traffic rate; finally, it is challenging to adapt LPL directly to newer radios like 802.15.4, since the specification limits the preamble size.

SCP-MAC adopts channel polling from LPL approaches. However, unlike LPL, SCP-MAC synchronizes the polling times of all neighbouring nodes. SCP-MAC distributes schedules much as developed by S-MAC: each node broadcast its schedule in a SYNC packet to its neighbours every synchronization period. The key is to discover the optimal synchronization period and wakeup tone length that minimizes the overall energy consumption. Also SCP-MAC eliminates long preambles, so its energy performance is not sensitive to varying traffics loads. The authors of SCP-MAC have implemented the protocols in TinyOS over the mica2 motes with the CC1000 radio, and to provide a clean comparison of LPL and scheduling, they implement SCP as a layer over basic LPL. They also describe a preliminary port to micaZ motes with the CC2420 radio supporting IEEE 802.15.4. The relative performance of SCP improves on never, faster radios like the CC2420, while that of LPL degrades.

Crankshaft [11] is a MAC protocol specifically targeted at dense wireless sensor networks. It employs node synchronization and offset wake-up schedules to combat the main cause of inefficiency in dense networks: overhearing by neighbouring nodes. Further energy savings are gained by using efficient channel polling and contention resolution techniques. Crankshaft employs a mechanism of channel polling very similar to the SCP-MAC.

3 Protocol Engineering

Protocol engineering becomes a key area of research enabling the development and deployment of power efficient networks. Cross-layer protocol design and power management have become two main approaches towards the development of such networks.

In order to explore the capabilities of state-of the-art technology, we have conducted a set of experimental trials [14]. The hardware equipment used throughout our experiments consisted of ten micaZ nodes (MOTE-KIT2400)[6] and a Stargate node used as base station. This latter node gathers the data collected from the ten sensor nodes. The nodes were located throughout different locations in a building characterized by different environmental conditions: sun light conditions, number of people normally working at a given lab, among others.

Our first experiment consisted in deploying them operating under the control of the application provided by the manufacturer. This applications collects and sends the data to the sink every eight seconds to the base station (high load scenario). We observe that the battery life span was 82 hours.

Due to the limited life span of the system, we did focus our research on the analysis of the power consumption, in particular, on the transmission system and associated protocols. It is well known that overhearing is the main source of power mismanagement. We choose two key elements to improve the lifetime of the network:

- To program a new application for data capture and data transmission to the base station
- To efficiently control the radio use. In a previous work, we have shown that the radio device consumes as much as 65% of the overall energy [12]

Regarding the second, we focus on the MAC layer and the different MAC protocols implemented for wireless sensor network: S-MAC, B-MAC or Wise-MAC [12]. All these protocols are different to the standard IEEE 802.15.4 [16]. They have important characteristics addressing the power consumption at the MAC layer of wireless sensor network. However, they have been developed for the mica2-based platforms (CC1000 radio chip) [17]. We have the started by adapting the B-MAC to the micaZ (CC2420 radio chip) [15].

B-MAC is able to reduce the *idle-listening*, i.e., the time that the node spends listening the channel. B-MAC requires that each node should wake up periodically to verify the channel activity. In case of detecting activity, the node keeps sensing the channel. On the contrary, no activity is detected, it falls asleep. The time between two consecutive wake-up periods is fixed by the check interval. B-MAC defines eight check intervals, each one corresponding to a different listening mode.

In order to ensure that all the packets are properly received, the packets are sent with a preamble whose length is longer than the check interval. B-MAC defines 8 sizes of preamble, each one related to a different way of transmission, which the protocol denominates transmit mode.

Another advantage of B-MAC, with respect to S-MAC, is its modularity and flexibility. B-MAC provides accessible interfaces to the upper layers, allowing the upper layers to transmit or listen in various modes.

B-MAC has been originally designed over the CC1000 radio chip (mica2). Therefore, the first task has been to analyze the main differences between the two radio chip systems. These are listed in Table 2.

Table 2. Summary of main sensor radio chip characteristics.

Chipcon CC1000	Chipcon CC2420
• Chip UHF RF transceiver	• 2.4 Ghz IEEE 802.15.4 systems
• Frecuency Range 300-1000 Mhz	• 250 Kbps
• Integrated bit synchroniser	• Suitable for both RFD and FFD
• Consumption: RX 9.6 mA TX 26.7 mA	• Consumption: Rx 18.8 mA TX 17.4 mA
	• Independent data buffering for rx and tx
	• Encrypt (AES 128)

As seen from the table, the use of the radio CC2420 provides some advantages to micaZ nodes, characteristics as throughput or the possibility to encrypt, but the fact that these new chip radio comply standards like IEEE 802.15.4, limits them in certain aspects. The preamble length is fixed by the standard. This is the main obstacle to the implementation of B-MAC on the CC2420 radio chip. B-MAC contains three important interfaces *MacControl*, *MacBackoff* and *LowPowerListening*, although the two early interfaces have already been implemented, the interface LowPowerListening (LPL) has been until today an insurmountable step.

Our proposal, according to the recommendations of the protocol designer, it is break with the idea of, if sender is not capable of prolonging the preamble for the total reception of the packet on the part of the receiver, that be the receiver that before the packet arrival event maintains the radio on for its total reception (see Figure 1). This is the main idea before the preamble problem, but will also be necessary to implement the rest of the interface that permit to level of application, and of transparent way, to define the listening and transmit mode desired.

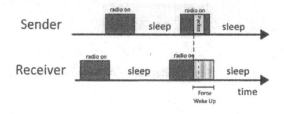

Fig. 1. On-off Timing.

The mechanism forcing the radio to keep awake should be incorporated into the LPL interface, i.e., made available to the upper levels. Our objective is testing the improvement using B-MAC and our own application, *IntellBuildApp*. We design different tests to evaluate the lifetime of network and to collect data allowing us to monitor the environmental conditions of the building. Four sets of trials were defined:

- Test 1: Use of the application as defined by the manufacturers based on the MTS400 and micaZ board. It sends a packet every 8 seconds, with all the data from sensor board. It makes use of the IEEE 802.15.4 MAC protocol.
- Test 2: Our own protocol. A packet is sent every 8 seconds, only temperature and humidity data are sent. It uses a multihop routing protocol and the standard 802.15.4 MAC protocol.
- Test 3: Same configuration as Test 2, but it uses the new interface implemented over B-MAC, using listening mode 3 and transmit mode 4 (1200 ms radio on, 1100 ms radio sleep).
- Test 4: Same configuration as Test 2, but it uses the new interface implemented over B-MAC, using listening mode 5 and transmit mode 3(2500 ms radio on, 900 ms radio sleep).

We evaluated the performance of the different set-ups in terms of the network lifetime and packet loss count. In this way, we can evaluate the energy-efficiency of B-MAC for wireless sensor networks. In terms of the network lifetime, the best results were obtained when the listening mode was longer than the transmit mode. Figure 2 graphically shows the results obtained. However, Figure 3 shows that the packet loss count is extremely high. This is due to the lack of synchronisation among the nodes. While some nodes may attempt to transmit, the potential receivers may be sleeping. It is clear that there is a need of a synchronisation scheme among the nodes. A

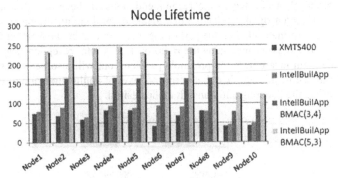

Fig. 2. Lifetime of different nodes

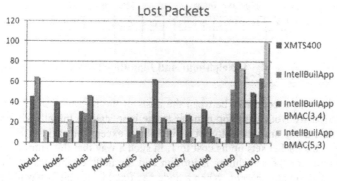

Fig. 3 Packet loss by node

4   SA-MAC: The Synchronous after Awake MAC engine

Bearing in mind the problem of packet loss and the need of developing an energy efficient protocol, we have undertaken the design of SA-MAC. We consider the design of a node synchronization engine. This synchronization algorithm would manage to save energy and establish routes, solving two issues in the area of wireless sensor networks. To achieve the design goal, we developed SA-MAC that consists of two phases: neighbour discovery phase and synchronization phase.

The main motivation towards the design of a synchronization scheme among neighbouring nodes is to limit the idle listen. Since the wake-up procedure implies a transient period, the synchronization should be designed by limiting the number of wake up steps. Towards this end, all the children nodes to a parent should send one after another before going back to sleep. In this way, the sink node only has to turn on its radio for a shortest period of time.

In our proposal, the nodes exchange their schedules by broadcasting them to all their immediate neighbours. It is also important to take into account the clock drift between different nodes since this is a critical point in synchronization

4.1 Neighbour Discovery Phase

In this first phase, each node sends periodic neighbour discovery packets by broadcasting them to all its immediate neighbours. In these packets, the node includes the time, it is important for establishing the listening period for the parent node. In our implementation, the discovery packets are sent every 20 seconds. If a node does not receive any discovery packet from a node for two minutes, it is locally considered inactive, i.e., no longer present.

When a node receives a neighbour discovery packet, it takes out the node id, its send time and cost to send through it and save it in its neighbour table, see Figure 4. Moreover, it adds an entry in its Planning Table.

Neighbour Table			Planning Table			
id	send time	active	id	cost	stage	delay
...	...	...	...	...	...	...

Fig. 4. Neighbour and Planning Tables.

For its election, it compares between the costs of the possible candidates, in this first evaluation we work with hop number to the base station. Initially all entries set up cost, stage, delay and active to not valid values, except for packets receive from base station, this packet sets up its cost to 1, in such a way that this entry has priority when is found for the scheduling algorithm and the node establishes it like parent setting up stage to value *parent*. When a node has fixed its parent, it has to inform to it for to be considered in the parent listen period. The node can also include the new cost to arrive to the base station in its discovery packet. In this way the network topology is generated as a tree, with the base station being the root with children nodes in a first level (hop) and the remaining ones in successive levels.

4.2 Synchronization Phase

In this second phase, we describe how the node chosen as parent synchronizes with all its children: As already stated, the children should send their packets one after the other, remaining the father the minimum time in the listen state.

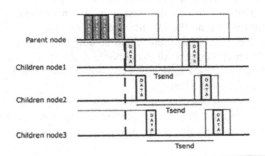

Fig. 5. Synchronization phase.

This task begins when a node receives a packet from another node. The parent sets up a record, in the Planning Table, for each child. In it, the parent indicates the time delay for the associated child. This time delay is fixed in accordance with the number of children associated to the father at that moment. We denote t_{ln} as the time that the parent node listens to each one of its children nodes and $N_{ch(i)}$ as the number of children of a parent at time i. This time is transmitted by the father to each child node.

$$delay_{node} = t_{ln} \times N_{ch(i)}$$

Once all children of a parent have received their delay, the parent sends a packet via broadcasting to synchronize all its children nodes. Figure 8 depicts the ideal situation of a parent node synchronized with three nodes.

It is a dynamic process in which new nodes can be easily integrated. A new node can be simply added to the end of the listening period. In this way SA-MAC is able to adapt the topology of the network due to the addition of a new node. In a similar way, nodes leaving the network can be simply discarded without requiring a long signalling process.

The following code shows the activity of node when a packet is received.

```
receive(packet)
{
    switch(packet_type)
    {
        case 'discovery':
            if (node_is_known)
                refresh_table();
            else insert_to_table();
            if(not_base_node)
                planning_algorithm();
        case 'information':
            if(node_is_known)
                insert_to_table();
            refresh_table();
            send_delay_to_children_node();
        case 'synchronization':
            save_delay();
        case 'start':
            send_event_to_application_layer_after_delay();
    }
}
```

5   Conclusions and Future works

This paper introduces the implementation of synchronized MAC for CC2420 radio chip. Energy efficiency is the primary goal in the protocol design. This is done by addressing the key features of the MAC layer, such as, overhearing and idle listening. By adopting a cross-layer design, we take advantage of the synchronization done at the physical level for developing a routing strategy. We are currently working on the development and testing of our proposal over an actual wireless sensor network.

References

[1] W. Ye, J. Heidemann and D. Estrin, An Energy-efficient MAC protocol for wireless sensor networks, In Proceedings of the 21st International Annual Joint Conference of the IEEE Computer and Communications Societies (INFOCOM 2002), June 2002

[2] W. Ye, J. Heidemann and D. Estrin, Medium access control with coordinated, adaptive sleeping for wireless sensor networks, In IEEE Transactions on Networking, April 2004.

[3] J. Polastre, A Unifying Link Abstraction for Wireless Sensor Networks, Doctoral Dissertation, University of California, Berkeley.October 25, 2005.

[4] J. Polastre, J. Hill and D. Culler, Versatile Low Power Media Access for Wireless Sensor Networks, In proceedings of the ACM SenSys'2004, November 2004, Baltimore, Maryland, USA

[5] T. van Dam and K. Langendoen, An adaptive Energy-Efficient MAC Protocol for Wireless Sensor Networks, In proceeding of the ACM Sensys'2003, November 2003, Los Angeles, California, USA

[6] I. Rhee, A. Warrier, M. Ahia and J. Min, Z-MAC: a Hybrid MAC for Wireless Sensor Networks, In proceedings of the 3rd International Conference on Embedded Networked Sensor Systems, 2005, San Diego, California

[7] S. Generiwal, D. Ganesan, H. Sim, V. Tsiatsis and M. Srivasta, Estimating Clock Uncertainty for Efficient Duty-Cycling in Sensor Networks, In proceedings of the ACM conference on Embedded Networked Sensor Systems, 2005, San Diego, California

[8] V. Raghunathan, S. Generiwal and M. Srivastava, Emerging Techniques for Long Lived Wireless Sensor Networks, IEEE Communications Magazine, April 2006

[9] M. Avvenuti, P. Corsini, P. Masci and A. Vecchio, Increasing the efficiency of preamble sampling protocols for wireless sensor networks, In proceedings of the 1st IEEE International conference on Mobile computing and Wireless Communications, Amman (Jordan), September 2006

[10] W. Ye, F. Silva and J. Heidemann, Ultra-Low Duty Cycle MAC with Scheduled Channel Polling, In proceedings of the ACM conference on Embedded Networked Sensor Systems, 2006, Boulder, Colorado, USA.

[11] G. P. Halkes and K.G. Langendoen, Crankshaft: An Energy-Efficient MAC-Protocol for Dense Wireless Sensor Networks, In proceedings of the 4th European Conference EWSN 2007, Delft, The Netherlands

[12] T.Olivares, P.J.Tirado, L. Orozco-Barbosa, V. López y P. Pedrón: Simulation of Power-aware Wireless Sensor Network Architectures, in ACM Internacional Workshop on Performance Monitoring, Measurement, & Evaluation of Heterogeneous Wireless and Wired Networks. Torremolinos, Málaga (España)

[13] K. Langendoen and G. Halkes, Embedded Systems Handbook, CRC Press, Aug. 2005

[14] T. Olivares, P.J. Tirado, F. Royo, J.C. Castillo and L. Orozco-Barbosa, Intell-Building: A Wireless Sensor Network for Intelligent Buildings, In proceedings of the 4th European Conference EWSN 2007, Delft, The Netherlands

[15] Crossbow Technology, INC MicaZ datasheet. www.xbow.com

[16] IEEE, Inc. Part 15.4: Wireless Medium Access Control (MAC) and Physical Layer (PHY) Specifications for Low-Rate Wireless Personal Area Networks (LR-WPANs), October 2003.

[17] Crossbow Technology, INC Mica2 datasheet. website. www.crossbow.com

Evaluating Energy Consumption of Proactive and Reactive Routing Protocols in a MANET*

Marco Fotino[1], Antonio Gozzi[1], Juan-Carlos Cano[2], Carlos Calafate[2], Floriano De Rango[1], Pietro Manzoni[2], and Salvatore Marano[1]

[1] University of Calabria
Pietro Bucci, 87036 Rende, ITALY
Email: {fotino, gozziantonio}@gmail.com
{derango, marano}@deis.unical.it
[2] Polytechnic University of Valencia
Camino de Vera, s/n, 46071 Valencia, SPAIN
Email: {jucano, calafate, pmanzoni}@disca.upv.es

Abstract. Ad hoc routing technology has been developed primarily for networks of mobile nodes. The operational life of untethered nodes will be limited by its power source, so ad hoc networks strongly depend on the efficient use of their batteries. All the layers of communication are coupled in power consumption and solutions addressing the power saving issue include transmission power control, power aware routing and low power modes at the physical layer. At the network layer, routing protocols may balance power consumption at nodes according to their routing decisions. Several authors have proposed power aware routing algorithms based on power cost functions. In this work we present a performance comparison of the DSR (Dynamic Source Routing) and the OLSR (Optimized Link State Routing) in terms of energy consumption. We evaluate how the different approaches affect the energy usage of mobile devices. We found that a reactive protocol takes advantage of its routing policy when the traffic load is low. However, at higher traffic rates, a proactive routing protocol can perform better with an appropriate refresh parameter. In addition, we showed how, independently from the routing protocol selected, the overhearing activity can seriously affect the performance since all the neighbours of a transmitting node also consume their energy. To the best of our knowledge, this is the first simulation study addressing the power saving issue to extensively compare the DSR and OLSR protocols under a wide variety of network scenarios.

1 Introduction

In the last few years, thanks to the proliferation of wireless devices, the use of mobile networks is growing very fast. In particular, a very large number of recent studies focuses on Mobile Ad-hoc Networks, also known as MANETs [1].

* This work was partially supported by the *Ministerio de Educación y Ciencia*, Spain, under Grant TIN2005-07705-C02-01. and by the *Generalitat Valenciana, Ayudas complementarias para proyectos de I+D+i*, Spain, under Grant ACOMP07/237.

A MANET is a network without a fixed infrastructure, in which every node can act as a router; this is required when the two end-points interchanging data are not directly within their radio range. This kind of network, self-organizing and self-reconfiguring, is very useful when it is not economically practical or physically possible to provide a wired networking infrastructure (battlefield scenarios, natural disasters, opportunistic networks etc.). Performance of a mobile ad hoc network depends heavily on the selected routing scheme, and the traditional Internet routing protocols do not work efficiently in a MANET. This kind of network, in fact, has a dynamic topology (every node can move randomly and the radio propagation conditions change rapidly over the time) and a limited bandwidth (so that the control traffic overhead must be reduced to the minimum) [2]. Developing routing protocols for MANETs has been an extensive research area in recent years, and many proactive and reactive protocols have been proposed from a variety of perspectives ([3]-[8]). These protocols try to satisfy various properties, like: distributed implementation, efficient bandwidth utilization, throughput optimization, fast route convergence and freedom from loops. Since mobile hosts today are powered by battery, efficient utilization of battery energy is a key factor. When a node exhausts its available energy, it ceases to function and the lack of mobile hosts can result in partitioning of the network, thereby affecting the overall communication performance.

The key to energy efficiency in future wireless terminals will be at the higher levels: low-energy protocols, energy-cognisant user interfaces, context dependent, and predictive shutdown management will be used to reduce the computation done at the terminal. The networked operation of a wireless terminal opens up additional techniques for increasing energy efficiency. Each protocol layer can use its own power conservation scheme. Electronically, we can turn off or slow down some devices (CPU, disk, antenna) when not needed, or adjust transmission power according to the interference level. At the data-link layer, energy conservation can be achieved using effective retransmission request schemes and powering off the nodes not participating in a communication in order to avoid the overhearing problem. At the network layer, we can modify routing metrics to prefer routes requiring lower levels of transmission power. These approaches can however affect other classical metrics such as the end-to-end delay and throughput, but it is a must to extend the network's lifetime.

In this work we measure and compare the energy consumption behaviour of two routing protocols: the Dynamic Source Routing (DSR) [9], which follows a reactive approach, and the Optimized Link State Routing (OLSR) [10], which uses a proactive one. We pursue a double objective. Firstly, we want to evaluate how different approaches affect the energy usage of mobile devices when using two of the most promising routing protocols currently considered under IETF's MANET working group [1]. In fact, among the great variety of different proposals, DSR and OLSR have arrive to the RFC status. Secondly, we want to check whether or not, under the IEEE 802.11 technology, some of the power aware routing proposals in the literature could be efficiently utilized to extend the lifetime of nodes and connections. In fact, we believe that, because of the overhearing and idle activity of a network interface card based on the current IEEE 802.11 technology, a majority of the proposed schemes not only are quite tricky to be implemented, but also could not achieve their assumed benefits. The simulation results presented in this paper were obtained using the ns-2 simulator [11], which is a discrete event, object oriented, simulator developed by the VINT project research group at the University of California at Berkeley.

This paper is organized as follows. In Section 2 we give a brief description of the routing protocols we compared, outlining their basic differences that could affect their energetic behavior. Section 3 describes the simulation environment and the energy model we implemented. Section 4 presents the obtained simulation results and finally, some concluding remarks are made in Section 5.

2   Routing protocols for MANETs

In this section we briefly review the main concepts dealing with the two protocols we analyzed, the DSR, and the OLSR respectively. This area is anyway a very dynamic one, and up-to-date information about the status of this area can be found in the web site of IETF's MANET group [1]. The MANET working group proposes two kinds of routing protocols: reactive and proactive. A reactive (or on-demand) routing protocol determines routes only when there is data to be sent. If a route is unknown the source node initiates a search to find one, and it is primarily interested in finding any route to the destination, not necessarily the optimal route. A proactive routing protocol, instead, attempts to continuously maintain routes to all destinations, regardless of whether they are required or not. To support this behaviour, the routing protocol propagates periodic information updates about network's topology or connectivity throughout the network. With respect to path selection, we can distinguish between source routing schemes, in which intermediate routers merely act as store-and-forward devices, sending the packet to the next hop according to the route indicated on the packet's header, and non-source routing systems. In the latter ones routers determine the path throughout the network based on their own calculations, assuming that hosts only have partial information about routes. Such routing protocols can be based on a link-state or a distance-vector algorithm. Traditionally, in a link-state algorithm each router floods routing information about the state of its own links to all nodes in the network, thereby building a picture of the entire network on its routing tables. On the other hand, in a distance-vector algorithm each router sends all (or a part) of its routing table only to its neighbors. From the addressing point of view, we can have a hierarchical routing system (some routers form a sort of backbone) or a flat address space (where all routers are peers of all others).

2.1   The DSR protocol

The Dynamic Source Routing protocol (DSR) is a reactive protocol which tries to reduce the overhead, while providing a reliable routing sheme at the cost of not finding optimal routes. Mobile hosts don't rely on periodic messages, with a consequently energetic advantage in terms of battery consumption. DSR only updates its routes when it needs to react to link failures on the routes being used. The protocol is based on the use of two main mechanisms: Route Discovery and Route Maintenance, which work together to allow nodes to discover and maintain routes to arbitrary destinations in the ad hoc network. The protocol allows multiple routes to any destination and allows each sender to select and control the routes used when performing routing decisions. The DSR protocol also includes a guaranteed loop-free routing and a very rapid recovery when routes in the network change. The DSR protocol has been mainly designed for mobile ad hoc networks of up to about two hundred nodes, and is designed to work well even at very high rates of mobility.

2.2   The OLSR protocol

The Optimized Link State Routing (OLSR) proactive protocol is an optimization of the classical link state algorithm, tailored to the requirements of a MANET. Because of their quick convergence, link state algorithms are somewhat less prone to routing loops than distance vector algorithms, but they require more CPU power and memory. Proactive protocols can be more expensive to implement and support but are generally more scalable. The key concept used in OLSR is the multipoint relay (MPR). MPRs are selected nodes which forward broadcast messages during the flooding process. This technique substantially reduces the message overhead as compared to a classical flooding mechanism where every node retransmits each message received. This way a mobile host can reduce battery consumption. In OLSR, link state information is generated only by nodes elected as MPRs. An MPR node may choose to report only links between itself and its MPR selectors. Hence, contrarily to the classical link state algorithm, partial link state information is distributed in the network. This information is then used for route calculation. OLSR provides optimal routes in terms of number of hops. The protocol is particularly suitable for large and dense networks as the technique of MPRs works well in this scenarios.

3   Simulation set-up

The simulation results presented in this paper were obtained using the ns-2 simulator. ns-2 is a discrete event, object oriented simulator developed by the VINT project research group at the University of California at Berkeley. The simulator has been extended to include: node mobility, a realistic physical layer that includes a radio propagation model, radio network interfaces and the IEEE 802.11 MAC protocol using the Distributed Coordination Function (DCF). The radio propagation model includes collisions, propagation delay and signal attenuation. In our experiments we have set a 54Mbps data rate, and a radio range of 250 meters.

3.1   Energy Consumption Model

A generic expression to calculate the energy required to transmit packet p is: $E(p) = i * v * t_p$ Joules, where: i is the current consumption, v is the voltage used, and t_p the time required to transmit the packet. We suppose that all mobile devices are equipped with IEEE 802.11g network interface cards (NICs). The energy consumption values were obtained by comparing commercial products with the experimental data reported in [12].

The values used for the voltage and the packet transmission time were: $v = 5V$ and $t_p = (\frac{p_h}{6*10^6} + \frac{p_d}{54*10^6})$ s, where p_h and p_d are the packet header and payload size in bits, respectively. We calculated the energy required to transmit and receive a packet p by using: $E_{tx}(p) = 280mA * v * t_p$ and $E_{rx}(p) = 240mA * v * t_p$, respectively. Since receiving a packet and just being idle, i.e., when simply powered on, are energetically similar [12], we assumed $E_{idle}(t) = 240mA * v * t$, where t is the NIC idle time.

Moreover, we account for energy spent by nodes overhearing packets. As shown in [12], we assume the energy consumption caused by overhearing data transmission is the same as that consumed by actually receiving the packet.

For the purpose of evaluating the effect of overhearing, we modified the energy model to account not only for the energy expenditure due to transmission and reception, but also for overhearing packet exchanges. Thus, the total amount of energy, $E(n_i)$, consumed at a node n_i is determined as:

$$E(n_i) = E_{tx}(n_i) + E_{rx}(n_i) + E_o(n_i), \tag{1}$$

where E_{tx}, E_{rx}, and E_o denote the amount of energy expenditure by transmission, reception, and overhearing of a packet, respectively. Notice that, as the average number of neighboring nodes affected by a transmission increases, the network is more dense, and so Eq.(1) implies that the packet overhearing causes much more energy consumption.

3.2   Methodology

To compare the DSR and the OLSR protocols, we simulated a dense wireless network, with 50 nodes moving in a 870×870 m area (with a density of about 66 nodes/km^2). Each node moves in this area according to the random waypoint mobility model, with a speed of 5 m/s and no pause time. In terms of traffic there are 12 CBR/UDP sources generating 20 packets/s, packet size is set to 512 bytes. The duration of each simulation is 450 seconds, with a startup period during the first 15 seconds where no traffic is generated.

Due to the random nature of the mobility model we used, the results of each simulation were considered as IID random variables (X1, X2, , Xn) with finite mean. We repeated the simulations, i.e., we varied to the value of n, to obtain an estimation of with a 95 percent confidence interval, by using the following definition:

$$\overline{X}(n) \pm t_{n-1,0.95}\sqrt{\frac{S^2(n)}{n}} \qquad (2)$$

where $t_{n-1,0.95}$ is the upper 0.95 critical point for Student's t distribution with n-1 degrees of freedom, $X(n)$ is the sample mean and $S^2(n)$ is the sample variance.

We mainly analyzed the time when each node dies due to lack of remaining battery (i.e., expiration time of nodes) as well as the lifetime of connection which captures the effects of disconnections due to lack of possible routes (i.e., expiration time of connections). We also measured the average end-to-end delay per packet, as well as the throughput. Finally, we also study how NIC activities contribute to the total energy expenditure. For the purpose of investigating the effect of overhearing, and according to the energy model described earlier, we modified the ns-2 energy model to allow measuring the battery energy consumed when overhearing packet exchanges, as well as the energy due to the idle operation mode.

4   Performance Study

4.1   Idle Power and Overhearing influence

We first evaluate the influence of the Idle Power on the energetic consumption of mobile nodes. To do that we assign different costs to the Idle state: zero, $\frac{E_{rx}}{4}$, $\frac{E_{rx}}{2}$, and E_{rx}. All nodes have their initial energy values randomly selected, but in a range that avoids nodes with extremely low levels of energy, which might not even attempt to start communication.

Figure 1 shows the results obtained. We observe that, only in case of no idle power consumption, can we clearly see the difference between the reactive an the proactive protocols being tested. In the other cases, even with a low idle state energy consumption, all the nodes in the network tend to exhaust their battery at about the same time (i.e. when idle power consumes all the device energy), no matter whether we are evaluating DSR or OLSR.

In addition, for the purpose of investigating the effect of overhearing, we modified the energy model to allow the battery power to be consumed by overhearing packets in the wireless channel.

Figure 2 shows how many nodes have died over time due to lack of battery. We plot the results with and without considering the effects of overhearing. We can definitely observe different results between two cases. When we consider overhearing, all approaches behave similarly, because neighbors of a transmitting node consume considerable part of their energy overhearing transmissions. Each node spends a very large amount of energy to overhear packets addressed to other nodes.

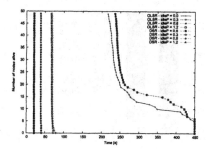

Fig. 1. Number of nodes alive vs time when varying idle power levels for the IEEE 802.11 techonolgy. $E_{rx} = 1.2mW$.

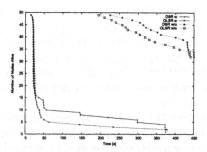

Fig. 2. Number of nodes alive vs time with (w) and without (w/o) considering overhearing.

We investigated the amount of energy consumed by participating nodes according to the network card activities. We observe that overhearing consumes most of the energy (see Figure 3). This implies that some techniques are required to reduce this energy expenditure by, for example, switching the network interface cards into the sleep mode.

Table 1 shows that the amount of energy spent in overhearing is larger than 90% for both protocols; which it mostly depends on node density and transmission range.

Table 1. Percentage of energy consumption: Transmission, Reception and Overhearing.

	DSR w/o	OLSR w/o	DSR w	OLSR w
Transmission	63.32	65.60	3.84	4.10
Reception	36.68	34.40	1.90	1.59
Overhearing	0.00	0.00	94.27	94.31

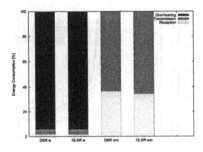

Fig. 3. Energy consumption(%) with(w) and without(w/o) considering overhearing.

We conclude that the idle power and overhearing effects dominate the energy consumption in the simulation of a dense network. Given this result, it seems that improvements should focus on devices' hardware to reduce the energy due to the idle state and to overhearing by, for example, switching the network interface cards into the off state. Otherwise, with the current technology, no significant differences will appear with respect to the energy consumption among reactive and proactive routing protocols.

We are well aware that IEEE 802.11 network interface cards, in the near future, could reduce the energy consumption due to the idle state. Furthermore, they could also include new modes with low cost in terms of energy consumption and transmission time, which could reduce the overhearing activities. So, in the next sections, we consider an ideal futuristic NIC where the impact of the idle state is negligible, and concentrate our study on the differences between reactive and proactive approaches. We make tests on two different scenarios: a completely static environment and a dynamic one such as the one presented in Section 3.2.

4.2 Static scenario

We first evaluate the behaviour of DSR and OLSR when all nodes maintain their initial positions throughout the simulation. The network consist of 27 mobile nodes distributed over a 550 × 750 meters area. In this scenario, 12 sender nodes establish a CBR/UDP connection with 12 receivers; data is relayed by three forwarding nodes situated in the center. Each CBR connection generates a constant bit rate of 4 packets/second with a packet size of 512 bytes.

The sender nodes will mostly consume energy due to transmission activities. The receivers will spend a larger amount of energy in receiving, while forwarders must perform both tasks, exhausting their energy earlier.

Figure 4 left shows lifetime of connection. When using the DSR protocol the three forwarders nodes exhaust their battery simultaneously, so that all the connections expire at about the same time. When using the OLSR protocol, due to its proactive behaviour, there is a small gap between expiration times. However, even though the proactive protocol spends energy to build and update routing tables during the setup period (the first 15 seconds of simulation), it is able to maintain the connections alive for about the same time as the DSR protocol.

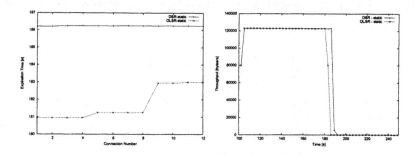

Fig. 4. Connections expiration (left) and throughput (right) in a static scenario.

Table 2 summarizes the performance of the protocols in the static scenario.

	DSR	OLSR
Data packet delivery ratio (%)	28.71	27.32
End-to-end delay (ms)	3.31	2.63
Overhead (%bytes)	0.34	12.09

Table 2. DSR - OLSR performance evaluation in a static scenario.

As for the end-to-end delay, it remains about 20% lower when using the OLSR protocol because the reactive DSR protocol must initially spend some time to find the path to the destination. With respect to the overhead, obviously it is higher when using the proactive approach. Finally, with respect to the data packet delivery ratio, the obtained results are almost the same independently of the selected protocol. The early exhaustion of the forwarding nodes' battery implies a network partition which seriously affects the packet delivery ratio. However, Figure 4 right shows that the throughput was maintained just until the network partition occurs.

4.3   Dynamic scenario

We now evaluate the impact of node mobility on performance. We simulated the DSR and the OLSR protocols in the dynamic scenario described in Section 3.2, and using the minimum hop count routing policy. Figure 5 shows the expiration time of nodes. We can observe how the DSR takes advantage from its reactive nature. During the first 15 seconds of simulation, while OLSR spends energy to update the network topology, DSR does not generate any packets. With respect to the lifetime of connections (see Figure 6 left) the response of OLSR and DSR is very similar, though obviously shifted since the proactive protocol starts its periodic exchange of message at the beginning of the simulation (15 seconds earlier).

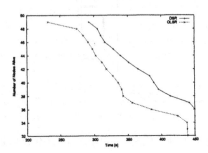

Fig. 5. Nodes alive vs time in a dynamic scenario.

Notice that, when compared to the static network, the lifetime of connections significantly increases. For the static network, some connections cannot progress when network partitioning occurs. However, for dynamic scenarios, node mobility favors the appearance of new paths after network partitioning disappears.

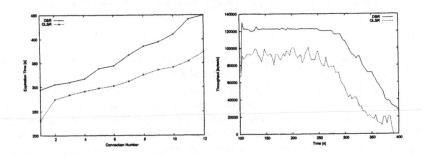

Fig. 6. Connections expiration (left) and throughput (right) in a dynamic scenario.

Table 3 summarizes the performance of the protocols in the dynamic scenario.

As expected, the DSR protocol performs better in terms of packet delivery ratio, end-to-end delay and routing overhead. The improvements in terms of routing overhead are expected since we are comparing reactive vs. proactive approaches. In terms of end-to-end delay notice that, with respect to the static

Table 3. DSR - OLSR performance evaluation in a dynamic scenario.

	DSR	OLSR
Packet delivery ratio (%)	92.68	71.15
End-to-end delay (ms)	12.23	13.54
Normalized routing overhead (%bytes)	0.86	14.77

scenario, we now have longer routes (on average). Concerning the packet delivery ratio, Figure 6 right shows detailed results for throughput variations throughout time. Before to the expiration of connections, DSR has a very stable throughput, while the one offered by the OLSR varies a lot. In a dynamic scenario, the reactive nature of the DSR protocol allows it to rapidly reacts to path changes. When using the OLSR protocol, node mobility leads to frequent packet losses. This effect could be reduced by updating the routing tables of OLSR more frequently, but this could lead to very high values of routing overhead.

Finally, Figure 7 shows the offered throughput as the number of nodes have died over time. As expected, as the number of nodes in the network decreases, some connections can not go ahead, and the total aggregated throughput is reduced. Again, the observed differences between the two protocols comes from the packet losses affecting the OLSR due to node mobility.

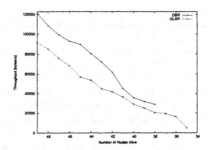

Fig. 7. Throughput vs number of nodes alive in a dynamic scenario.

5   Conclusions

In this work we analyze the energy consumption behaviour of the Dynamic Source Routing protocol and the Optimized Link State Routing protocol. We evaluate how different models for power consumption affect the energy usage of mobile devices. We found that a reactive protocol takes advantage from its routing policy when the traffic load is low. However, at higher traffic rates, a proactive routing protocol can perform well with an appropriate refresh parameter.

Extensive simulations under different MANET scenarios showed the importance of considering the overhearing and the idle mode as part of the energy model. We observe that, with the current technology, the idle power and overhearing effects dominate the energy consumption, and so new techniques should be investigated to reduce the energy consumption during idle periods due to overhearing.

The obtained results induce us to believe that some of the solutions found in the literature based on power-aware route selection metrics can not achieve their assumed benefits using real hardware. Future work will focus on new approaches to extend the lifetime of both nodes and connections by combining the sleeping mode with new power-aware mechanisms.

References

1. Internet Engineering Task Force, "Manet working group charter", http://www.ietf.org/html.charters/manet-charter.html.
2. C.-K. Toh "Maximum Battery Life Routing to Support Ubiquitous Mobile Computing in Wireless Ad Hoc Networks", in IEEE Communication Magazine, June 2001
3. E.Royer and C.-K. Toh, "A Review of Current Routing Protocols for Ad Hoc Mobile Wireless Networks", IEEE Personal Communications Magazine, Vol. 6, No. 2, April 1999.
4. C-K. Toh, "Associativity Based Routing For Ad Hoc Mobile Networks", Wireless Personal Comm. Journal, Special Issue on Mobile Networking and Computing Systems, Vol. 4, No. 2, Mar. 1997.
5. Z.J. Haas, "A New Routing Protocol for the Reconfigurable Wireless Networks", Proceedings of the ICUPC'97, San Diego, CA, Oct. 1997.
6. V.D. Park, M.S. Corson, "A Highly Adaptive Distributed Routing Algorithm for Mobile Wireless Networks", IEEE INFOCOM'97, Kobe, Japan, 1997.
7. C.E. Perkins, E.M. Royer, "Ad-hoc On-Demand Distance Vector Routing", Proceedings of the 2nd IEEE Workshop on Mobile Computing Systems and Applications, Feb. 1999.
8. D. Dube, C.D. Rais, K-Y. Wang, K. Tripathi, "Signal Stability-Based Adaptive Routing (SSA) for Ad hoc Mobile Networks", IEEE Personal Comm., pp. 36-45, Feb.,1997.
9. David B. Johnson, David A.Maltz, Yih-Chun Hu, and Jorjeta G. Jetcheva, "The Dynamic Source Routing Protocol for Mobile Ad Hoc Networks DSR)", Internet Draft, February 2002.
10. P.Jacquet, P.Muhlethaler, T.Clausen, A.Laouiti, A.Qayyum, L.Viennot, "Optimized Link State Routing Protocol for Ad Hoc Networks"
11. K. Fall and K. Varadhan, (Eds.) "ns Notes and Documents", The VINT Project. UC Berkeley, LBL, USC/ISI, and Xerox PARC, February 25, 2000. Available from http://www.isi.edu/ salehi/ns_doc/.
12. Laura Feeney and M. Nilsson, "Investigating the Energy Consumption of a Wireless Network Interface in an Ad Hoc Networking Environment," IEEE INFOCOM, 2001.

Ultra-low Power Sensors with Near Field Communication for Mobile Applications

Esko Strömmer[1], Mika Hillukkala[1], Arto Ylisaukko-oja[1]

[1] VTT, P.O. Box 1100, FI-90571 Oulu, Finland

E-mail: Firstname.Surname@vtt.fi

Abstract. In this paper, we study the applicability of an emerging RFID-based communication technology, NFC (Near Field Communication), to ultra-low power wireless sensors. We present potential application examples of passive and semi-passive NFC-enabled sensors. We compare their NFC-based implementations to corresponding implementations based on short-range radios such as Bluetooth and Wibree. The comparison addresses both technical properties and usability. We also introduce Smart NFC Interface, which is our general-purpose prototype platform for rapid prototyping of NFC and Bluetooth implementations. Two pilot sensor implementations and an NFC-Bluetooth gateway implementation based on our platform are presented and evaluated. Finally, needs and possibilities for technical improvements of available NFC technology are discussed.

Keywords: Near field communication (NFC), Mobile applications, Wireless sensors

1 Introduction

Sensors with inductive data transmission and powering have been under research activities for several years. Examples can be found in [1], [2] and [3]. Proposed solutions have usually been based on proprietary technologies or standard RFID technologies. A problem concerning commercial mobile sensor applications has been the lack of standardized inductive technology in mobile handsets such as mobile phones and PDAs, which has made a specific reader device or add-on module mandatory.

In March 2004, a new interconnection technology, Near Field Communication (NFC), was launched by Sony, Philips and Nokia, by the establishment of the NFC Forum. The NFC Forum is a non-profit industry association for advancing the use of NFC short-range wireless interaction in consumer electronics, mobile devices and PCs. In co-operation with ECMA, the NFC Forum will promote the implementation and standardization of NFC technology to ensure interoperability between devices and services. As of the end of 2006, about 100 members have joined the NFC Forum. Within the next few years, NFC is forecast to penetrate commercial mobile phones [4].

We expect NFC to provide a potential solution for enhancing existing and emerging mobile applications with data acquisition from various sensors. NFC

Please use the following format when citing this chapter:

Strömmer, E., Hillukkala, M., Ylisaukko-oja, A., 2007, in IFIP International Federation for Information Processing, Volume 248, Wireless Sensor and Actor Networks, eds. L. Orozco-Barbosa, Olivares, T., Casado, R., Bermudez, A., (Boston: Springer), pp. 131-142.

supports the use of mobile handsets by touch-based interaction, which is an intuitive and user-friendly way of establishing connections and exchanging information between mobile handsets and other devices [5]. From the sensor application viewpoint, NFC can provide easy-to-use touch-based access to the sensor data by low-cost mobile phones and other mobile handsets, which will make the application specific reader device unnecessary and thus decrease the system level costs dramatically. As NFC is capable of transferring power between devices, fundamentally NFC enables semi-passive implementation of sensors with multi-month to multi-year battery lifetime and even passive implementation without any power source. However, this calls for commercial NFC technology that supports zero-power operation of the NFC transceiver (similar to an RFID tag) when waiting for activation from an active NFC device, and efficient power management during communication.

2 Near Field Communication Technology

Near field communication (NFC) is a very short-range (max. 20 cm, but typically only a few cm), wireless point-to-point interconnection technology, evolved from a combination of earlier RFID contactless identification and interconnection technologies (ISO14443A/MIFARE/FeliCa). It enables users of handheld electronic devices to access content and services in an intuitive way by simply "touching" smart objects (e.g. sensors, RFID tags, other handheld devices), i.e. connecting devices just by holding them next to each other. The communication is based on inductive coupling. The 13.56 MHz carrier frequency is used and the available data rates are 106, 212 and 424 kbps. The related standards are shown in Fig. 1.

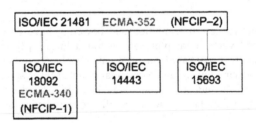

Fig. 1. NFC-related standards. The upper layer defines the mechanism of selecting the communication mode on the lower layer [6].

The legacy of earlier standards gives NFC compatibility benefits with existing RFID applications, such as access control or public transport ticketing – it is often possible to operate with old infrastructure, even if the RFID card or reader is replaced with an NFC-enabled mobile phone, for example. This is possible because of NFC's capability to emulate both RFID readers ("reader/writer mode") and RFID tags ("card emulation mode"). NFC hardware can include a secure element for improved security in critical applications such as payments. For example, a credit card could be integrated into a mobile phone and used by contactless credit card readers over NFC.

In addition to the reader/writer and card emulation modes, there is an NFC-specific NFCIP-1 mode ("peer-to-peer mode"), defined in the ECMA-340 standard. This mode is intended for peer-to-peer data communication between devices. In this mode, NFC is comparable to other short-range communication technologies such as Bluetooth, Wibree and IrDA, although the physical data transfer mechanism is different. In this respect, NFC can be seen as a rival of these technologies, even though it can also complement them. NFC can open a connection between two devices that are brought close to each other, and the actual communication will then occur by a longer range technology such as Bluetooth.

In peer-to-peer mode, the participant that starts the communication is called the initiator and the other participant the target. The peer-to-peer mode is divided into two variants: active mode and passive mode. In active mode, both participants generate their own carrier while transmitting data. In passive mode, only the initiator generates a carrier during communications, and the target device uses load modulation when communicating back to the initiator, in a way similar to passive RFID tag behavior [7]. This makes it possible to save power in the target device, which is a useful feature if the target device has a very restricted energy source, such as a small battery. Fundamentally it is possible to make a target device operate for several years without battery replacement or even without any battery by the power that is emitted by the initiator.

NFC is a standard technology that has recently achieved commercial availability via NFC chips, modules, mobile phones and PDAs. NFC is also backed by the leading mobile phone manufacturers and its deployment and chip development will be strongly driven via its integration into cellular handsets. According to the downgraded forecast of the NFC-enabled mobile handsets by ABI Research in September 2006, the number of shipped NFC-enabled handsets in 2011 will be 450 million units, or 30 per cent of all handsets.

The most natural choice for NFC-enabled sensors is the passive peer-to-peer (NFCIP-1) communication mode. Also the card emulation mode is feasible.

3 Application Scenarios

In this chapter we present some application scenarios of sensors with NFC connectivity. We point out that these are not yet commercial products but only examples of the versatile application possibilities of NFC-enabled sensors. Our scenarios are divided into three main groups:

- passive (also batteryless) sensors,
- user controlled semi-passive (also semi-active or battery assisted) sensors, and
- stand-alone semi-passive sensors for long-term monitoring.

3.1 Passive Sensors

A fundamental operation principle of RFID systems is that the passive end of the communication link (tag) gets all its energy for operation from the field generated by the active end of the communication link (reader). It is also possible to apply this principle in the implementation of NFC-enabled passive sensors by replacing the RFID reader with an NFC device that operates in reader-writer or in peer-to-peer initiator mode, and the tag by an NFC-enabled sensor that operates in card emulation or in peer-to-peer passive target mode. We have estimated that at least several milliwatts of power can be delivered to an NFC antenna that has a diameter of 25 mm with field strengths defined in the associated NFC standard (1.5 - 7.5 A/m). With larger antennas, the delivered energy can be higher. If the transferred power is not enough for the sensor due to, e.g. a smaller antenna or longer range, it is possible to add a powering capacitor into the sensor module that is charged before the measurement and from which power is taken during the measurement. Because battery replacement or wires are not required, these sensors can be easily embedded in various structures (e.g. walls, injection molded products) during their manufacturing.

An example of a low power sensor that could easily be made NFC-enabled is a silicon based temperature sensor such as LM20 [8]. The current consumption of this sensor is only 10 μA, thus only a minor part of the typical power transferred over an NFC link. An A/D-converter and some logic are needed between the temperature sensor and the NFC transceiver. This can be implemented separately or included in the same microcontroller that takes care of the NFC protocol. Another example is a moisture sensor that could be embedded into walls and flooring of the house close to water pipes or floor drains for indicating possible leaks that usually cannot be detected outside the structure until they have caused severe damages to the structure.

3.2 User-Controlled Semi-passive Sensors

Semi-passive (also semi-active or battery assisted) RFID tags have an on-board power source (for example, a battery) and electronics for performing specialized tasks. The on-board power supply provides energy to the tag for its operation. However, for transmitting its data, a semi-passive tag uses power that is emitted by the reader in the same manner as passive tags [9]. Instead of a battery, a parasitic power supply exploiting ambient energy such as temperature gradients, light, mechanical forces and movements or electromagnetic fields from other sources than the reader are possible.

The same operation principle could also be applied to NFC-enabled sensors, in which case we call them semi-passive sensors. User-controlled semi-passive sensors are in a non-operating state most of the time, a state in which their current drain from the on-board battery is negligible. The sensor is activated by some action from the user. This can be pushing a switch-on button or bringing an active NFC-enabled device to the vicinity of the sensor, in which case its NFC transceiver that operates in card emulation or in peer-to-peer passive target mode activates the entire sensor.

User-controlled semi-passive sensors could be well used, for example, in health or disease management in which monitoring is based on sparse measurements (e.g. blood pressure, weight). In these kinds of applications, the amount of data is typically limited. The sensor is switched on and the measurement activated by the user. Measurement results could be transferred to an NFC-enabled mobile phone either immediately or later in blocks of several results acquired over time. Scenarios of this kind are discussed more closely in [10]. Another potential application could be a battery assisted implanted blood glucose sensor such as in [11] with activation and communication by NFC instead of proprietary radio technology. However, due to the very limited size of implant sensors and thus also very small maximum antenna size, the technical feasibility of NFC-enabled implant sensors requires further investigation, and the maximum communication range will in any case be much shorter than in the other examples mentioned above.

3.3 Stand-Alone Semi-passive Sensors for Long-Term Monitoring

Semi-passive sensors could also be applied to various autonomous long-term monitoring applications. In addition to user actions, these sensors have to be able to be activated by other external events or periodically by an internal real-time clock or other wakeup timer. This will make them a little more complicated than the user controlled sensors, but still technically feasible.

As an example, an NFC-enabled condition monitoring sensor could be fitted onto goods that require special care during transport. At the reception end, the sensor is checked by an NFC-enabled reader device that gives an alarm if something unexpected has occurred during the transport. Another application area could be long-term physical activity and health monitoring. For example, an inertial sensor integrated into a shoe could measure the daily walking distance, and the result could be read from the sensor at the end of the day by an NFC-enabled mobile phone. A potential application example is also an RFID tag size electric supply meter that could be integrated into any electrical appliance and could take its supply power from its mains supply capacitively. Again, the cumulated energy consumption of the appliance could be read by an NFC-enabled mobile phone.

4   NFC vs. Short-Range Radios

The nature of NFC-based wireless sensors clearly differs from many conventional implementations of wireless sensors and sensor networks. In short-range radio based wireless sensor networks, there is not necessarily a human user at all, but the sensors may serve as a part of an automation system, for example. In contrast, a human user plays a central role in typical NFC-based implementations, a mobile handset such as a mobile phone being the typical tool for the interaction.

Bluetooth enabled phones are currently widely available. However, according to [4], also NFC should become a common technology in mobile phones in the near future. Comparing these technologies is thus relevant. Fundamentally the NFC communication offers the following advantages over Bluetooth:

- NFC enables intuitive, easy-to-use touch-based communication and interaction between two devices.
- Communication set-up latency with NFC is typically some hundreds of milliseconds, whereas with Bluetooth it is typically several seconds.
- NFC enables longer lifetime of the sensor battery, or even batteryless implementation of the sensor.
- Pure NFC communication enables lower price, since NFC is technically less complex than Bluetooth.
- Due to its shorter range and near field coupling, NFC is more immune to eavesdropping and intentional or unintentional interferences.

The very short communication range of NFC could be seen as a mere shortcoming, but the matter is not that simple. From the user point-of-view, the association between communicating devices is very intuitive with NFC: to start the communication, just "touch" a device with the mobile handset, instead of trying to identify the correct device from a potentially long menu in the mobile handset, as in the case of Bluetooth communication. This type of physical browsing user paradigm may prove to be a very powerful one from the human-computer interaction perspective [5]. In an ideal situation, no key presses or browsing of menus is required, or at least this is kept to a minimum. For example, confirmation by a key press can sometimes be necessary for security reasons.

Still, communication range is the main limitation of NFC compared to Bluetooth. This, depending on the application, can create following disadvantages:

- NFC does not suit portable devices that require online connectivity to another portable device or to a fixed access point.
- Lower bit rate together with the short communication range can make the touch-based transfer of longer data blocks unpleasant.
- The placement of the antenna is more critical. The location of the antenna has to be indicated to the user.

These disadvantages can be partly overcome by combining NFC with Bluetooth or WLAN, which on the other hand will mean that some of the advantages such as the lower price of pure NFC implementation are lost. Still, this is an important aspect in NFC application possibilities. In semi-passive applications requiring transmissions of longer data blocks, NFC may be used only to initiate the connection, and the actual data transmission is then performed by means of Bluetooth or WLAN which offer higher data rates and longer communication range. In that case, NFC is a tool to configure the connection easily and explicitly with a simple touch, instead of configuring the connection by performing device searches or setting of configuration parameters by browsing menus and using keypads. Compared to pure NFC, the user does not need to keep the devices close to each other after the connection has first been initiated. In a similar way, NFC can also be used to configure long-term device pairs (such as Bluetooth trusted pairs).

A classic problem with wireless sensors is the source of power. A reasonable solution is essential for the overall usability of the system. Otherwise the maintenance of the system may require too much effort. Good examples of such cases are the implantable and wall-embedded sensors mentioned in Chapter 3, but this easily becomes an issue even in far more conventional scenarios. The trend is to decrease the power consumption of radios to enable longer operation time from small batteries (Bluetooth derived Wibree technology, for example [12]). However, this only makes the maintenance interval longer instead of eliminating the need for maintenance completely. Basically the only way to avoid maintenance for energy supply in case of short-range radios would be to apply solutions like harvesting power from ambient energy sources such as light, temperature gradients, vibration and so on. Unfortunately, this method tends to increase complexity and price, and requires very case-specific designs.

Here NFC (or RFID) based implementations have the potential of being superior to other currently available technology alternatives: even fully passive implementations are possible with some NFC chips such as Microread by Inside Contactless [13], getting the required energy for both the measurement (and/or a secure element) and communications from the reading device. This enables potentially maintenance-free devices.

If there is no energy storage, the restriction is that the measurement can only take place while the reader is nearby. In some applications (see Chapter 3.3), a semi-passive implementation is needed instead to enable measurements also at other times. Even then, the average current consumption of an NFC-enabled semi-passive sensor can be on the order of tens of microamperes without sacrificing the communication latency – which typically needs to be sacrificed when optimizing power in the case of other short-range radio technologies.

5 Pilot Demonstrators

5.1 Smart NFC Interface

Smart NFC Interface is a multi-purpose platform that we have developed for evaluating the NFC technology, rapid prototyping and demonstrating of related product ideas, and fastening the associated product development by technology transfer. It is a light, matchbox-sized device with a microcontroller, rechargeable Li-Ion battery with charging electronics, data logging memory, RS-232 serial port and other wired communications, as well as NFC, IrDA and Bluetooth wireless communications. The structure is modular to facilitate modifiability for different applications. There are connectors for extensions such as sensors. An LM20 temperature sensor [8] is readily included, and the design supports ultra-low power operating modes. However, passive (batteryless) sensors are not yet supported. The microcontroller is programmable by C. A photograph of the electronics and the basic application scenario in sensor applications is shown in Fig. 2.

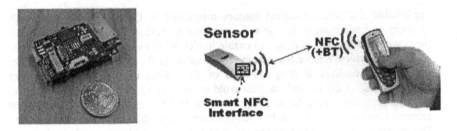

Fig. 2. Smart NFC Interface module (Communication Board on top, NFC antenna not installed) and its basic application scenario as a sensor interfacing module.

The block diagram of the Smart NFC Interface electronics is shown in Fig. 3. In its basic form, the device contains two boards: 1) Basic Board, and 2) Communication Board. The dimensions of both boards are 56 x 31 mm. Overall thickness of the device including the default rechargeable battery is 16 mm without encapsulation (boards attached one on top of another).

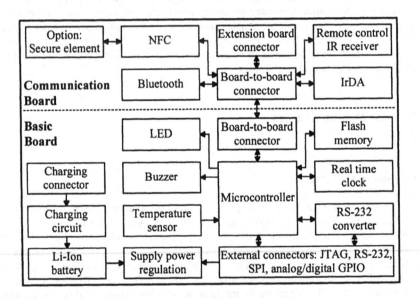

Fig. 3. Block diagram of the Smart NFC Interface electronics

The Smart NFC Interface has a single 8-bit microcontroller (Atmel AVR series ATmega128L). The microcontroller is equipped with 128 kilobytes of flash program memory that enables easy reprogramming and debugging via a JTAG interface, 4 kilobytes of internal RAM and 4 kilobytes of EEPROM. An external 8 Mbit flash memory is connected to the SPI bus of the microcontroller. This can be used for data logging purposes in stand-alone semi-passive sensors (see Chapter 3.3), for example.

The Smart NFC Interface is equipped with a Philips PCF8563T real-time clock (RTC) circuit, whose main purposes are 1) enabling recording of timestamps for events, and 2) serving as a wakeup timer, thus facilitating ultra-low power modes. Both of these are particularly useful features in stand-alone semi-passive sensor implementations for long-term monitoring according to Chapter 3.3.

A versatile power supply circuitry is used, enabling power to be taken not only from the internal battery, but also from RS-232 serial port or other external supplies. The power supply automatically adjusts to a wide range of input voltages. The charging circuit is connected to the LED, which indicates the charging status. While not charging, the LED is also software controlled by the microcontroller. Another supported UI feature is a buzzer.

On the Communication Board, the NFC transceiver is based on NXP Semiconductor PN531 chip, which is connected to the microcontroller via its SPI bus. A SmartMX secure chip can optionally be installed. Bluetooth (Class 2) is based on Bluegiga WT12-A-AI module, which contains Bluetooth protocol layers up to RFCOMM and profiles, and is connected to the microcontroller by UART. IrDA support currently exists only in hardware, based on Microchip MCP2150-I/SO IrDA controller and Zilog ZHX1403 IrDA transceiver. There is also hardware support for a remote control infrared (IR) receiver.

The general purpose IO (GPIO) on the external connector includes an analog input. This can be used to connect sensors. Digital interfaces such as RS-232/UART and SPI are also included. RS-232 levels are converted to logic level UART and vice versa by Maxim MAX3222CAP.

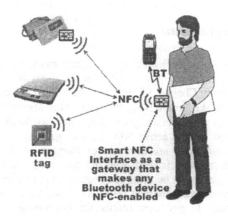

Fig. 4. Applying the Smart NFC Interface as a gateway

Two pilot NFC sensor implementations based on the Smart NFC Interface are presented in Chapters 5.2 and 5.3. Since the NFC-enabled phones and their JSR-257 contactless communication API standard do not currently support peer-to-peer communication, we have also configured the Smart NFC Interface to operate as a gateway device between non-NFC mobile phones and NFC-enabled sensors (see Fig. 4). The gateway is an active NFC device in peer-to-peer initiator or reader/writer mode and is connected to a host device (a mobile phone or a laptop) by Bluetooth. When the gateway detects a target NFC device or an RFID tag, it opens the NFC communication, reads the data from the target and sends the data to the host device.

5.2 Temperature sensor

The temperature sensor pilot is an example of a user controlled semi-passive sensor according to Chapter 3.2 that is activated by another NFC device. The pilot consists of a Smart NFC Interface module with an inbuilt temperature sensor and an active NFC-enabled mobile device. This can be either an NFC-enabled mobile phone or a Smart NFC Interface module that is configured to gateway mode (see the previous chapter).

At the sensor end, the Smart NFC Interface sets its NFC transceiver to target mode and waits for an active NFC device. When it is detected, the inbuilt temperature sensor is powered up and after a short stabilization time the temperature sensor output is sampled. The result is converted into Celsius degrees and sent by the NFC transceiver as an ASCII character string.

The gateway sets its NFC transceiver to poll target NFC devices in initiator mode. After detecting the target temperature sensor, it sends a command via the NFC communication link to it and waits for a reply. After receiving the ASCII character string from the temperature sensor, it sends it to the host device via Bluetooth for displaying it to the user.

5.3 Energy consumption meter

Another pilot that we have implemented is a slightly modified version of the previous example, being an example of a stand-alone semi-passive sensor for long-term monitoring according to Chapter 3.3. In this pilot, a separate sensor box is connected to the serial port of the Smart NFC Interface. This sensor box monitors electric current inside a mains wire with a tiny coil that is attached on the surface of the wire.

The program flow on the active NFC device (gateway) side is the same as in previous example, but there are a couple of changes on the target side. After a measurement result message is read from the sensor box to the target Smart NFC Interface module, there has to be some sanity checks to make sure that the message is not corrupted. Then the measurement result is extracted from the message structure, calibrated and added to previous results for calculating the cumulated energy consumption. The message to be sent through NFC is constructed by making use of this newest measurement result.

All this data processing takes quite a long time and it is not possible to do these calculations during NFC communication, since the NFC initiator would timeout before all calculations would be finished on the NFC target (sensor) side. Therefore calculations must be carried out before the NFC communication. When an active NFC device is detected, the pre-calculated data is sent to it directly from the memory of the target NFC device.

The current consumption of the Smart NFC Interface operating in target mode (sensor) is approximately 10 mA for the host processor and 25 mA for PN531. It is possible to reduce the average total current consumption (host + PN531) to 3.5 mA by periodically starting and stopping PN531, and using the RTC to wakeup the host processor from sleep every 50 milliseconds. According to PN531 datasheet [14] it is possible to put PN531 into power down mode where its current consumption is only 30 µA. In this mode, the combined current consumption during sleep periods would be 40 µA, and PN531 should be able to detect an active NFC device and wake up the host processor by using its interrupt line. Unfortunately this power down mode does not work with the SPI interface of PN531 that we are using in our design.

The current consumption of the gateway is approximately 10 mA for host processor and 90 mA for the NFC chip PN531. With antenna tuning, it is possible to affect the PN531 current consumption, but it will also have an effect on the reading distance. It is also possible to use NFC in short bursts, for example three read cycles per second, reducing the average total current consumption to 15 mA without sacrificing the short latency time evidently.

The NFC communication data rate in our pilots is 106 kbps. It takes a few tens of milliseconds to open an NFC communication channel and send a few bytes of measurement data. This delay is very short compared to delays produced in mobile phone software and user interface.

6 Discussion

We conclude that NFC is a high-potential technology for implementing ultra-low power sensors with touch-based communication with mobile handsets. Contrary to other RFID technologies, NFC is backed by the leading mobile phone manufacturers and is expected to penetrate into mobile phones in near future. This will improve the availability of NFC technology in general and facilitate its applicability in mobile sensor applications, too.

However, from the wireless sensor perspective, there are still limitations in commercially available NFC chip and phone implementations. Thus we were not able to pilot all application scenarios that we presented in Chapter 3. For example, the PN531 by NXP Semiconductors supports passive RFID tag type of load modulation in NFC communications, but it does not support taking energy for all operations from the reader side. This prevents implementing passive sensors based on this device. On the other hand, the lowest power consumption of the device with receiver active is 30 µA from 2.5 volts [14]. This is still quite a lot and does not enable coin-cell battery lifetime of several years in semi-passive implementations. We also failed to utilize the full potential of the device, based on giving a wakeup interrupt for an external microcontroller when an NFC initiator comes nearby. This caused a severe compromise to power optimization.

NFC chips are also being offered and developed to support fully passive operation. The main reason for this trend is the target to enable critical NFC functionality, such as access control and payments in mobile phones, even if the battery has been completely discharged. An example of such a device is described in [13]. NFC chips having suitable interfaces, support for passive operation and a high degree of integration are also needed for making the implementation of NFC-enabled sensors easier. Good power management support is essential for semi-passive sensors. NFC chips with reduced functionality (e.g. without active modes) could be a reasonable approach to ultra-low power sensors, since this would cut down their power consumption and price further.

The JSR-257 API standard for NFC phones needs to be evolved further to include support for peer-to-peer mode. The range of NFC-enabled mobile handsets is still limited, concerning especially smartphones. This can be partially overcome by an NFC-Bluetooth gateway that we have demonstrated. This kind of approach of having a separate, wireless NFC communication device may be desirable even in the long run in some special applications.

References

1. Li, Y.L., Liu, J.: A 13.56 MHz RFID Transponder front-end with Merged Load Modulation and Voltage Doubler-clamping Rectifier Circuits. Circuits and Systems. IEEE International Symposium (2005) 5:5095-5098.
2. Chevalerias O., O'Donnell T., Power D., O'Donovan N., Duffy G., Grant G., O'Mathuna S. C.: Inductive Telemetry of Multiple Sensor Modules. IEEE Pervasive Computing, Vol. 4 No. 1 (2005) 46-52
3. Sawan M.: Wireless smart implants dedicated to multichannel monitoring and microstimulation. Proceedings of the IEEE/ACS international Conference on Pervasive Services (2004) 21-26
4. Near Field Communications (NFC). Simplifying and Expanding Contactless Commerce, Connectivity, and Content. ABI Research, Oyster Bay, NY (2005)
5. Välkkynen, P. Korhonen, I. Plomp, J., Tuomisto, T., Cluimans, L., Ailisto, H., Seppä, H.: A user interaction paradigm for physical browsing and near-object control based on tags. Proceedings of Physical Interaction Workshop on Real World User Interfaces, Udine,IT, 8-11 Sept., 2003. University of Udine, HCI Lab., Department of Mathematics and Computer Science (2003), 31 - 34
6. ECMA International web site (2003). http://www.ecma-international.org/publications/standards/Ecma-352.htm.
7. ECMA International standard ECMA-340. 2nd Edition (2004). http://www.ecma-international.org/publications/files/ECMA-ST/Ecma-340.pdf.
8. National Semiconductor LM20 web site (2007). http://www.national.com/pf/LM/LM20.html.
9. Lahiri S.: RFID: A Technology Overview. IBM Press (2005). http://www.ibmpressbooks.com/articles/article.asp?p=413662&rl=1.
10. Strömmer E., Kaartinen J., Pärkkä J., Ylisaukko-oja A., Korhonen I.: Application of Near Field Communication for Health Monitoring in Daily life. Proceedings of 28th IEEE EMBS International Conference. New York City, USA (2006) 3246-3249.
11. SMSI Glucose Sensor. Sensors for Medicine and Science, Inc. (2007). http://www.s4ms.com/products_glucose.htm.
12. Wibree technology data sheet. http://www.wibree.com/technology/Wibree_2Pager.pdf
13. Inside Contactless Microread fact sheet. http://www.insidecontactless.com/products/microread-eng.pdf
14. NXP Semiconductors PN531-µC based Transmission module Objective Short Form Specification February 2004 Revision 2.0.

Modelling QoS for Wireless Sensor Networks

José-F Martínez[1], Ana-B García[1], Iván Corredor[1], Lourdes López[1], Vicente Hernández[1] and Antonio Dasilva[1]

[1] EUIT Telecomunicación - DIATEL
Universidad Politécnica de Madrid
Ctra. Valencia, Km. 7, 28031, Madrid, SPAIN
{jfmartin, abgarcia, icorred, llopez, vhernandez, adasilva }@diatel.upm.es

Abstract: A wireless sensor network (WSN) is a wireless network composed of spatially distributed and tiny autonomous nodes – smart dust sensors, motes –, which cooperatively monitor physical or environmental conditions. Nowadays these kinds of networks support a wide range of applications, such as target tracking, security, environmental control, habitat monitoring, source detection, source localization, vehicular and traffic monitoring, health monitoring, building and industrial monitoring, etc. Generally, these applications have strong and strict requirements for end-to-end delaying and loosing during data transmissions. In this paper, we propose a realistic scenario for application of the WSN field in order to illustrate selection of an appropriate approach for guaranteeing performance in a WSN-deployed application. The methodology we have used includes four major phases: 1) Requirements analysis of the application scenario; 2) QoS modeling in different layers of the communications protocol stack and selection of more suitable QoS protocols and mechanisms; 3) Definition of a simulation model based on an application scenario, to which we applied the protocols and mechanisms selected in the phase 2; and 4) Validation of decisions by means of simulation and analysis of results. This work has been partially financed by the "Universidad Politécnica de Madrid" and the "Comunidad de Madrid" in the framework of the project CRISAL - M0700204174.

Keywords: Wireless Sensor Networks (WSN), QoS protocols, performance, target tracking, natural environments surveillance.

1 Introduction

Recently, we have witnessed significant evolution in the field of wireless sensors. The latest stage has been characterized by improvements in sensor hardware issues (miniaturization of pieces, increased ROM and RAM capacities, more energy capacity, etc). These facts and the new field of possibilitiess for their application have boosted interest in Wireless Sensor Networks (WSN). WSN might be defined as follows: *Networks of tiny, small, battery-powered, resource-constrained devices equipped with a CPU, sensors and transceivers embedded in a physical environment where they operate unattendedly.* While a good deal of research and development has been carried out in architecture and protocol design, energy saving and location, only

Please use the following format when citing this chapter:

Martínez, J.-F., García, A.-B., Corredor, I., López, L., Hernández, V., Dasilva, A., 2007, in IFIP International Federation for Information Processing, Volume 248, Wireless Sensor and Actor Networks, eds. L. Orozco-Barbosa, Olivares, T., Casado, R., Bermudez, A., (Boston: Springer), pp. 143-154.

a few studies have been done on network performance in WSN (Quality of Service – QoS).

The service provided by the network is closely related to the quality of that service. Traditional QoS requirements (usually from multimedia applications) such as bounded delays or bandwidth are not pertinent when applications are tolerant of latency or the size of the packets being transmitted is very small. Generally, packet delivery ratio is an insufficient metric in WSNs: what is important is the amount and quality of information that can be extracted from a WSN.

Some studies on QoS have focused on protocols and mechanisms for MAC and the network layer, and almost all these have been developed and tested through simulations. All these approaches for supporting QoS in WSN can constitute a base for future work in this direction, and they obviously represent the starting point in our proposal. We have already conducted work on state-of-the-art QoS in WSNs. This work has focused mainly on QoS-based protocols and mechanisms both in MAC and network layers. The results of this work can be consulted in [1].

The remainder of the paper is organized as follows:

The case study is depicted in section 2. In this section the proposed application scenario is described, as are all its QoS-related characteristics. Based on these characteristics, we have identified the QoS mechanisms which are needed both in MAC and network layers of the protocol stack. The section concludes with a selection of the most suitable protocols for MAC and network layers available in literature on WSN. The validity of decisions on QoS protocols and mechanisms is verified in section 3, with the use of simulation software to perform tests. Previously, we have designed a simulation model to which we have applied the protocols selected in section 2. Section 4 concludes this paper with an overview of future research activities.

2   Case study: QoS in forest fire detection

In this section we will apply the study we have presented on QoS protocols in WSN [1] to a forest surveillance scenario. So we will begin by extracting the QoS-related requirements from the real-time forest surveillance application, allow us to select the network and MAC protocols later that best suit these requirements. However, these protocols may not meet all necessary requirements. If so, we will also propose add-on features for each protocol. We will also create a simulation model from the application and subject it to a number of simulation tests. In the conclusions section, we will discuss what we see as the shortcomings of the protocols studied herein, and which one should be corrected in future research.

2.1   Description and analysis of requirements of application for real-time forest surveillance

The application will focus on both forest fire detection and event tracking in a natural environment (natural reserve) of great ecological importance. The main objective of the application will be the early detection of forest fires to avoid ecological disasters. Likewise, the application will have secondary objectives such as the detection and tracking of intruders within protected spaces for the prevention of illegal actions. In short, the application will be used for forest surveillance, including detection of dangerous activities and conditions that increase the risk of fires; detection and location of fires; fire monitoring and assistance in fire extinction; detection and tracking of intruders entering restricted areas.

In our forest fire detection application, sensor nodes collect measurement data, such as relative humidity, temperature, infrared radiation, COx and NOx gases. Other components of the WSN supporting our application are laptops and/or PDAs (as support to firemen and safety watchmen), a server and a data base. All WSN services will be accessible to remote users through web services. Figure 1 illustrates the proposed scenario.

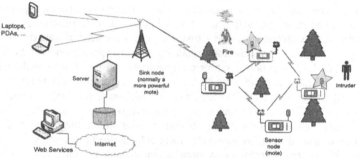

Fig 1. Forest surveillance application scenario

All sensors will be used to determine the risk of fire at a given moment. The infrared radiation sensor will also be used for the detection and tracking of intruders in restricted areas. Specifically, the application will have the following characteristics:

1) Topology and network dynamics: The WSN topology is a design parameter that should be taken into account when guaranteeing QoS. The selected topology for the WSN will be flat. Therefore, every node will have the same hierarchy in the WSN as well as the same hardware components. The hierarchy will not be necessary in the proposed network since it will use a localized geographic routing.

2) Geographical information: Sensor nodes must obtain geographical information – i.e., coordinates – in order to locate the events within the natural reserve. Methods commonly used to acquire this data are based on GPS [2] or distributed location services [3]. For WSNs, a GPS-based approach is too expensive, thus our WSN implements a distributed location service. However, this method adds certain overheads during the initial phase of the WSN that could impede ensuring QoS at those moments.

3) Real-time requirements: Fire monitoring or target tracking reflects the physical status of dynamically changing environments such as temperatures or positions of moving targets in forest areas. This sensory data is valid only for a limited time; hence it needs to be delivered within a time deadline.

4) Unbalanced mixture traffic: Another characteristic which will considerably affect QoS decisions is *reactive-proactive* hybrid behaviour. Reactive behaviour will come from fire or intruder detection, and will generate traffic to the sink node according to the event-driven delivery model. This traffic type is generated aperiodically through the detection of critical events at unpredictable points in time. Proactive behaviour will come from the monitoring of the environmental status and tracking targets, and will generate traffic to the sink node according to continuous delivery model.

5) Data redundancy: High redundancy in the sensor data is a common characteristic to most WSNs. Redundancy may improve several QoS requirements, such as reliability and the robustness of data delivery. However, this uses a large amount of energy. To solve this problem, we could use data fusion or data aggregation to maintain robustness while decreasing redundancy in the data, but these mechanisms require high levels of computational activity in at least several nodes (usually cluster-heads). Therefore, these mechanisms also add delay and complicate QoS design in WSNs. We prefer to exclude these mechanisms, as our application is based on two critical objectives, and real-time requirements will prevail over energy requirements. An alternative to data aggregation and fusion is the meta-data negotiation which is able to eliminate redundancy without introducing excessive delay in data delivery.

6) Energy efficiency: An important challenge to this application will be energy efficiency. The large number of sensor nodes involved in the WSN and the need to operate over a long period of time (from 6 months to 1 year) will require careful management of energy resources. However, to implement the QoS mechanism to support critical real-time traffic and while saving energy is not a minor task. The key is to distribute the energy load among all sensor nodes so that the energy at a single sensor node or a small set of sensor nodes will not be drained too quickly. Nowadays, achieving this energy distribution without compromising the QoS requirements is very difficult since mechanisms and protocols do not usually consider both possibilities at the same time.

7) Sensor data priority: Not all sensing data are equal; hence they have different levels of importance. For example, the data generated in a fire detection event will have more importance than that generated in the monitoring for determining the conditions that increase the risk of fire. QoS mechanisms will determine the data delivery priorities for the different data types existing in the WSN.

As a result, QoS support for the network will take into account almost all of the aforementioned characteristics in the application specifications. The next section describes how to extract network and MAC layers mechanisms from QoS-related requirements of the protocol stack according to the application characteristics analyzed.

2.1.1 Network layer

Guaranteeing network layer QoS for diverse traffic types is a challenge, as WSN characteristics such as dynamic topology change as a result of a number of factors: node failure, addition or mobility, the large scale of a network with thousands of densely-placed nodes, periodical and aperiodical traffic generated by sensors with different priorities and real-time requirements, or possible data redundancy produced by correlated sensor nodes.

Traditional network layer methods based on end-to-end path discovery, resource reservation along the path discovered and path recovery in case of topological changes will not be suitable for our WSN: initially, the time wasted in the path discovery is not acceptable for urgent aperiodic – i.e., event-driven - packets. Moreover, it is not advisable to reserve resources for unpredictable aperiodic packets. Even for periodic continuous flows, these methods are not practical in a dynamic

WSN because service disruption during path recovery increases data delivery delay, which is not acceptable in our mission-critical application. Finally, end-to-end path-based approaches are not scalable due to excessive overheads related to path discovery and recovery in large scale sensor networks. As an alternative to inefficient reservation-based approaches, the network layer will include an end-to-end QoS provisioning method based on local decisions at each intermediate node without path discovery and maintenance.

To solve dynamic topology changes, the network layer will implement the aforementioned localized geographic routing. This type of routing will mainly provide adaptability to dynamic topology changes, since the nodes will not require acquisition of global topology information. Consequently, no control packet will be generated in significant amounts with topology changes due to node addition, failure or mobility. The nodes in the WSN will able to take localized packet routing decisions without a global network state update or a priori path setup, thus increasing network scalability and decreasing the control traffic. Further, this routing scheme is suitable for both critical aperiodic and periodic packets as a result of no path setup and recovery latency.

Another characteristic that should be included in the network layer is traffic priorities. In our WSN, the traffic priority will be characterized by two domains: reliability and timeless. The network layer will implement complex mechanisms in order to achieve this objective. For example, it could implement a priority queue system for the purpose of differentiating among traffic with different end-to-end deadlines. On the other hand, the mechanisms that will be implemented for lending reliability to data transmissions could exploit the inherent multiple redundant paths to the final destination in a dense WSN to guarantee the required end-to-end level of reliability (end-to-end reaching probability) of a packet. Finally, the network layer will not implement a mechanism for eliminating data redundancy such as data aggregation, for the aforementioned reasons. Alternatively, the network protocol will implement a method for dealing with redundant data by exchanging meta-data in so-called data negotiation [4]. This eliminates the inefficiencies generated by data-aggregation mechanisms resulting from flooding and the subsequent processing of information. For instance, if a tracking event is detected and a data negotiation mechanism is used, location information is transmitted once and no further data is transmitted until the target moves.

2.1.2 MAC layer

Not all of the aforementioned QoS requirements could be met by network layer. Consequently, our WSN protocol stack will have a MAC protocol capable of performing the following tasks: medium access control according to packet deadlines, measurement of the average delay to individual neighbours, the measurement of the rate of loss to individual neighbours. In addition to, it may be necessary to have the capacity to deliver the packet to multiple neighbours reliably.

Along with the aforementioned functionalities, the MAC layer must implement mechanisms where each one of deadlines assigned by network layer is associated to a transmission priority level. Thus, medium access prioritization will be achieved through the MAC layer. Likewise, the MAC protocol will be able to measure the average delay to individual neighbours with the purpose of forwarding the packet according to its deadline.

However, packet forwarding will be performed not only on the basis of deadline criteria but also on those of reliability. For this reason, the MAC protocol will measure the rate of loss to individual neighbours.

The localized geographic routing used by the network layer will require transmission of control packets with the position data of neighbours situated at least one or two hops away. For the transmission of these control packets, the MAC layer will require the capacity to reliably deliver multicast packets.

2.1.3...Selected network and MAC protocols
Considering the mechanisms just described, the following design decisions have been made for the network and MAC protocols in our surveillance application:

From network layer perspective, we believe that only a few of the protocols of surveyed in [1] could be used in our WSN. We have selected three candidates from among these protocols: MMSPEED [5], SPEED [6] and Directed Diffusion [7]. (See table 1).

Table 1. Comparative table of routing protocols in Wireless Sensor Networks.

	Network topology	Data delivery model	Data aggregation/fusion	Traffic guarantees	Several traffic classes	Networks dynamics	Resources reservation	Scalability
Directed Diffusion	Flat	Query-driven and Event-driven	Yes	Reliability	No	Limited	Yes	Medium
SPEED	Flat	Query-driven and Event-driven	No	Soft Real-time	No	No	No	Low
MMSPEED	Flat	Event-driven and Continuous	No	Reliability and Real-time	Yes	Limited	No	High

We selected these protocols for several reasons:

MMSPEED
→MMSPEED implements localized geographic routing, which is fundamental for the network layer of our stack protocol. These mechanisms increase self-adaptability of the network to dynamic changes as well as scalability of the network. In addition, this protocol is suited for both periodic (real-time) and aperiodic traffic because routing decisions are local (i.e., no path setup and failure recovery).

→MMSPEED also implements a multi-speed mechanism to assign diverse deadlines to the packets with different delay requirements. This mechanism is ideal for supporting multiple traffic types (continuous, event-driven, etc.). Its dynamic speed compensation mechanism, which is capable of immediately correcting small inaccuracies produced in initial routing decisions, is also quite useful.

→Routing decisions in MMSPEED are also made according to the reliability level required by the packet. To route on the basis of the reliability requisite, MMSPEED has an advanced method of lending reliability to data transmissions which involves using the frame loss rate of the MAC layer to make an estimate of the reliability level of each link.

However, MMSPEED lacks a method for dealing with the data redundancy problem. We have already mentioned that the best methods for eliminating data redundancy in our application are those based on meta-data exchange. In this sense, we are in the course of studying how a meta-data negotiation mechanism can be added to MMSPEED.

SPEED

→SPEED is another QoS routing protocol for WSN that provides light real-time end-to end guarantees. The QoS mechanism, which is employed by SPEED, is based on a distance to sink estimate for guaranteeing fulfillment of delay requirements. This can be an useful mechanism for prior traffic in our WSN.

→The network layer will accept the packet depending on the required speed. If the QoS mechanism verifies that it will not be capable of achieving the delay requirement for a certain packet, then the packet will be discarded before it is forwarded to a neighbor node.

→SPEED can be recovered by means of back-pressure mechanism if the network becomes congested. This feature can be decisive for ensuring that data is transported to the sink with an acceptable delay.

→Like MMSPEED, SPEED bases its routing decisions on geographic localization of sensor nodes. This routing mechanism can notably increase network scalability. The routing module in SPEED (SNGF) implements a distributed database where a node can be selected in order to attain the speed requirement.

SPEED also has several disadvantages for our WSN. First, SPEED treats all traffic classes equally. However, as we already commented, there will be at least two traffic classes (aperiodic and periodic traffic) in our WSN that will require different QoS levels in terms of reliability and delay. Secondly, SPEED does not implement any reliability mechanism, which may be necessary for critical traffic.

Directed diffusion

→ Directed Diffusion is a data-centric and application-aware paradigm. This protocol implements a mechanism based on data aggregation to eliminate redundant data coming from different sources. This particularity reduces the number of transmissions drastically, leading to two main consequences: firstly, the network saves energy and extends its life-time, and secondly, it has higher bandwidth in the links near to the sink node.

→Directed diffusion is based on a query-driven model. This means that the sink node requests data by means of broadcasting *interests*. When events begin to appear, they start to flow towards the originators of interests along multiple paths. This behavior provides reliability and robustness to data transmissions in the network.

Although Directed Diffusion includes all these optimization mechanisms, the protocol has two shortcomings in the realm of QoS: directed diffusion can neither explicitly manage QoS parameters such as delay and reliability, nor differently handle more than one traffic class.

For the MAC layer, we have established the following criteria:

First, selecting a MAC protocol that complements MMSPEED protocol is no secondary decision. MMSPEED specifications propose an extension of 802.11e for supporting all mechanisms implemented by the network layer. The most important of these is the priorities mechanism. However, this MAC protocol is not specific to WSNs and consequently bears some deficiencies as a result. We propose the Z-MAC [8] protocol as an alternative to 802.11e. Although this protocol needs several additional features to be completely compatible with MMSPEED, it is an excellent starting point because it implements a priority mechanism that is very appropriate for this case study. The additional features are mainly concerning hybrid nature of Z-MAC. The latter forces the priority mechanism to work in a different way, depending on its contention level (low level - CSMA or high level TDMA). In addition, Z-MAC must be capable of associating each MMSPEED's speed layer with a priority class in the MAC layer.

On the other hand, Z-MAC has a highly efficient contention method that can avoid unnecessary backoff delays in packet transmissions. Another distinctive feature of Z-MAC is its adaptability to topology changes.

Another MAC protocol that can be used in our WSN is B-MAC [9]. B-MAC is not a QoS-specific protocol but it does include several interesting mechanisms that can notably optimize the WSN. The features that best define B-MAC are its simplicity of design and implementation, in addition to its flexibility, allowing it to offer multiple classes of service and to adapt to any scenario.

A comparative analysis between Z-MAC and B-MAC is shown in the following table.

Table 2. Comparative table of MAC protocols in WSN.

	Data aggregation/fusion	Scalability	Priority mechanisms	Energy aware	Contention-based
B-MAC	No	High	No	Yes	Yes
Z-MAC	No	High	Yes	Yes	hybrid

3 Simulation of application scenario

3.1 Simulation model

The following table depicts the simplified simulation model defined for the application described in section 3.1:

Table 3. Simulation Environment Settings.

Size terrain	600mx600m
Terrain morphology	A mountain of 400mx400m, centered in the terrain.
Sensor node number	176 nodes (sink included)
Radio range	80 m
Initial energy charge	1000 Joules
Bandwidth	200 Kbps
Payload	32 bytes

The sensor nodes are deployed around the mountain, distributed in four sectors (*North, South, West* and *East*). The sink node is placed at coordinate (0,0).

J-SIM is simulation software selected to implement the model. It was chosen because it is component-based, a feature that enables users to modify or improve it.

Network protocols have been configured with different parameters according to capacities. All the parameters defined for each protocol are depicted in following sub-sections:

MMSPEED

Table 4. MMSPEED parameters.

	Attaining sink probability	Max. delay (in seconds)
High priority traffic (events)	0.4	0.5
Low priority traffic (monitoring)	0.2	4

Moreover, we have defined two speed layers which have been configured with different speed levels (1000 m/s and 250 m/s, respectively).

SPEED

SPEED has been configured to ensure a delay of 0.8 seconds in all packet transmissions. Besides, the transmission speed will not have to exceed a value of 1000 m/s. These parameters are applicable to all packet types.

Directed Diffusion

Directed Diffusion can be configured with multiple parameters. The most significant parameters for the simulation tests are the following: diffusion area of interests (complete area); duration of interests (all time simulation); interest refresh (every 10 seconds).

3.2 Simulation results

Since J-SIM can only simulate the MAC protocol IEEE 802.11 (in all its variants), the results analyzed in this section are for the behavior of QoS mechanisms in network layer. At present, our work group is considering the possibility of integrating Z-MAC and B-MAC inside of J-SIM simulator.

Deadlines

-**MMSPEED**: The results of simulations with MMSPEED are significant in the way they show how the protocol is capable of differentiating traffic classes (see Figure 2).

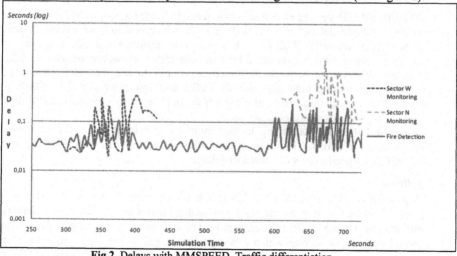

Fig 2. Delays with MMSPEED. Traffic differentiation.

When low and high-priority traffic concurs in the WSN, MMSPEED successfully supported the QoS level assigned to both traffic classes. The maximum delay configured for high-priority traffic (0.5 seconds) was never exceeded. Furthermore, the jitter (or delay fluctuation) is not excessively high, which will improve the quality of real-time data received by the application, especially if these data have been generated by the tracking of a person inside the area monitored by the WSN. In addition, low-priority traffic manages to maintain acceptable levels of delay, although the jitters are somewhat high. This fact will not lead to a decline in the quality of data obtained from monitoring, as they are not real-time data (they are generally stored in the database for later enquiries).

Figure 3 shows delays recorded in monitoring of traffic in a simulation period using SPEED and Directed Diffusion protocols. This graphic shows only a specific example, but it can be extrapolated to the complete simulation time.

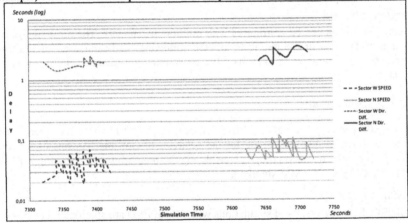

Fig 3. Delays with SPEED and Dir. Diff. Comparative graphics.

-SPEED: SPEED showed excellent performance in handling both traffic classes. The packets are never delayed more of 0.19 seconds, which is below the maximum limit allowed (0.8 seconds). SPEED also manages to maintain the jitters within an acceptable range, which is positive for real-time traffic. However, because SPEED does not differentiate between traffic classes it has to use the same amount of resources for routing both high priority traffic and low priority traffic, which is unnecessary. Any increase in monitoring traffic could seriously compromise real-time traffic.

-Directed Diffusion: According to the results, it is evident that the mechanisms implemented in Directed Diffusion are insufficient to ensure the QoS level required by the WSN, specifically with regard to delays.

Reliability

-MMSPEED: When MMSPEED initiates a packet flow to the sink following a period of inactivity, it is common for intermediate nodes to have incoherent routing information. Until MMSPEED recovers operational status, tenths of seconds to one second may elapse, during which a few packets might be discarded. However, the discarding of packets does not mean an effective loss of event notifications, as there is always a route whose nodes have information on the correct routing, and these can consequently route the packets towards the sink. In other cases, MMSPEED shows a great robustness due to its multi-path mechanism.

-SPEED: SPEED has a significant lack of reliability because it does not implement any mechanisms to guarantee that packets reach the sink node. This characteristic of SPEED means that when congestion occurs, all discarded packets will inevitably become lost packets. On the other hand, SPEED does not undergo periods of instability caused routing information inconsistencies.

-Directed Diffusion: The simulation tests with Directed Diffusion were satisfactory in terms of reliability. Although Directed Diffusion does not implement an explicit mechanism to provide reliability, it achieves an acceptable reliability level by means of a multi-path routing that selects the best paths towards the sink.

Energy consumption

During the first 12 hours of simulated time, the consumption of energy of the eight nodes closest to the sink was recorded. The results can be seen in the table 4.

Table 5. Energy consumption with each protocol.

	Average energy consumption	Lifetime in a real WSN
MMSPEED	3.5225 Joules/hour	9 months
SPEED	6.5056 Joules/hour	5.6 months
Directed Diffusion	0.9575 Joules/hour	3 years

The first column shows the average levels of energy consumption in the simulation period. The second column shows the lifetime of a real WSN, assuming that sensor nodes use AA alkaline batteries. It is evident that the results vary greatly.

Directed Diffusion showed the best rate of energy consumption (3 years aprox.). These good results have been obtained through use of the data aggregation mechanism implemented by Directed Diffusion. This mechanism significantly reduces the number of data transmissions, and therefore helps saves a great deal of energy.

MMSPEED achieves an acceptable lifetime (9 months) and a very good energy/delay balance. However, this lifetime could be increased if MMSPEED implemented a mechanism for reduction of redundant data (e.g. meta-data negotiation or data aggregation).

SPEED has showed the worst results in energy consumption (6 months). This is mainly due to the large number of control packets transmitted by SPEED. SPEED may be insufficiently scalable to be used in a large WSN.

Taking into account all the simulation results, we conclude that the most suitable protocol for improving performance in our WSN is MMSPEED. However, this protocol should be improved with several add-on features, which should be the subject of future research. In section below, we discuss a number of improvements that could be made to the protocol.

4   Conclusions and future work

In this paper we have presented a study of MAC and network layer protocols that have been defined to provide QoS in wireless sensor networks. We have focused on the basic mechanisms used in these protocols for guaranteeing performance parameters to applications, leading to charts comparing the different approaches.

Taking this study as a basis, we have also selected a forest surveillance application in order to show how appropriate protocols for QoS could be selected by defining the performance requirements of the application and the classification criteria for the protocol study.

This research has also shown what we consider to be shortcomings in the protocols. For instance, the MMSPEED protocol lacks a data aggregation or an even more preferable meta-data negotiation system. Other aspects that could be considered in more detail in MMSPEED are the energy-delay trade-off, and the facility for parameter interchange with MAC layer. As for the Z-MAC protocol, the initial overhead, the prioritization system and its lack of a data fusion mechanism are examples of issues that could be improved. In future work, we intend to modify Z-MAC to achieve full compatibility with MMSPEED.

We are presently working on defining and subsequent deploying a WSN scenario in which a surveillance application will be run. For future research, and after the functional aspects of the application are working, we plan to include performance monitoring in the system. This will allow us to perform empirical studies of the influence of the parameters we have considered on the quality offered to the application. At the moment, we are performing simulation experiments using J-SIM [10], which we have modified for our case study. However, its adaptation to MAC protocols is still poor, and we are working to solve this problem.

5 References

1.   J.F. Martínez, A.B. García, I. Corredor, L. López, V. Hernández and A. Dasilva, "QoS in Wireless Sensor Network: Survey and Approach". To be published in Proc. IEEE/ACM EATIS, May, 2007

2.   B. Karp and H. Kung, "Greedy Perimeter Stateless Routing for Wireless Networks", Proc. IEEE/ACM Int'l Conf. Mobile Computing and Networking, pp 243-254, 2000.

3.   T. He, C. Huang, B. Blum, J. Stankovic, and T.Abdelzaher, "Range-Free Localization Schemes for Large Scale Sensor Networks", Proc. Mobicom Conf. , 2003.

4.   J. Kulik, W. R. Heinzelman, and H. Balakrishnan, "Negotiation-based protocols for disseminating information in wireless sensor networks," Wireless Networks, Volume: 8, pp. 169-185, 2002.

5.   E. Felemban; Chang-Gun Lee, Ekici, E. MMSPEED: multipath Multi-SPEED protocol for QoS guarantee of reliability and. Timeliness in wireless sensor networks. Mobile Computing, IEEE Transactions on Volume 5, Issue 6, pages 738-754, June, 2006.

6.   T. He, J. Stankovic, L. Chenyang, and T. Abdelzaher. SPEED: A stateless protocol for real-time communication in sensor networks. In Proceedings of 23rd International Conference on Distributed Computing Systems, pages 46–55, May, 2003.

7.   C. Intanagonwiwat et al., "Directed diffusion: A scalable and robust communication paradigm for sensor networks", in the Proc. of MobiCom'00, Boston, MA, August 2000.

8.   Z-MAC: a Hybrid MAC for Wireless Sensor Networks; Injong Rhee, Ajit Warrier, Mahesh Aia and Jeongki Min (Technical Report, Department of Computer Science, North Carolina State University, April 2005).

9.   J. Polastre, J. Hill, and D. Culler. Versatile low power media access for wireless sensor networks. In SenSys '04: Proceedings of the 2nd international conference on Embedded networked sensor systems, pages 95--107, New York, NY, USA, 2004. ACM Press.

10.   Sobeih, A.; Hou, J.C.; Lu-Chuan Kung. "J-Sim: a simulation and emulation environment for wireless sensor networks". Wireless Communications, IEEE. Volume 13, Issue 4, Aug. 2006 Page(s):104 – 119.

Wireless Sensor Network Localization
using Hexagonal Intersection

Eva M. García, Aurelio Bermúdez, Rafael Casado, and Francisco J. Quiles

Instituto de Investigación en Informática de Albacete (I^3A)
Universidad de Castilla-La Mancha
02071 Albacete, Spain
{evamaria.garcia, aurelio.bermudez, rafael.casado, francisco.quiles}@uclm.es

Abstract. Device localization or positioning is a key issue in wireless sensor networks applications. One solution widely used to estimate the position of a node in the network consists in using the intersection of coverage areas. For the sake of simplicity, these areas have traditionally been modeled by rectangles, assuming some extra inaccuracy. In this paper we propose a localization algorithm based on hexagonal intersection. Results show that the only substitution of the geometric shape provides better results in the localization of the devices composing the network.

Keywords: Wireless sensor networks, localization.

1 Introduction

A wireless sensor network (WSN) consists of a set of electronic devices able to collect physical information about the environment where they have been disseminated, to process that information, and to transmit it through a wireless medium. Each device (popularly called *mote*) incorporates a microcontroller (integrating a processor and a memory), a radio transmitter and receiver, a battery, and a sensor card, which gives name to the set, and allows to measure a great variety of physical magnitudes (pressure, temperature, humidity, etc.).

Application scope of WSNs is greater every day. These networks are used in fields such as precision agriculture, environmental monitoring and forecasting, military applications, and natural disasters. Practically, all these applications require that each sensor knows their accurate location in the space. In many cases, it is enough to have an approximation.

Usually, it is difficult, maybe impossible, to program this position at each device in advance. In outdoor environments, it is possible to use a satellite positioning system, like GPS (*Global Positioning System*) [5]. However, nowadays it is an unfeasible option to equip every node in network with a GPS receiver, essentially due to size, energy consumption, and price reasons. Hence, other techniques have been gained popularity. These algorithms assume that a small number of devices, so called *anchors* or *beacons*, already know their location. These nodes propagate this

Please use the following format when citing this chapter:

García, E. M., Bermúdez, A., Casado, R., Quiles, F. J., 2007, in IFIP International Federation for Information Processing, Volume 248, Wireless Sensor and Actor Networks, eds. L. Orozco-Barbosa, Olivares, T., Casado, R., Bermudez, A., (Boston: Springer), pp. 155-166.

information to the rest of the network. Then, the other nodes can infer their position through a distributed algorithm.

Increasing the mathematical complexity of these algorithms carries advantages and disadvantages. On the one hand, a parallel increment in the accuracy of the estimate is produced. On the other hand, the battery consumption is increased during the process, because of the executed mathematical operations and the amount of performed communications.

A subset of the existing localization algorithms can be classified according to the kind of geometric figure used to represent the possible localization area. Thus, algorithms based on circular intersection are located on one end, providing great accuracy, but requiring the use of trigonometric functions and complex data structures. On the other end, we have algorithms based on rectangular intersection, with very low computational cost and more imprecision in the estimate.

In this work we propose one solution establishing a tradeoff between both points: hexagonal intersection. The use of hexagons provides more accuracy on devices localization. Moreover, extra complexity does not represent an important obstacle, particularly if we consider new generation motes, with higher performance with regard to computation capacity and power consumption.

The rest of this paper is organized as follows: Section 2 presents taxonomy of main kinds of localization algorithms, and focuses on solutions using rectangular intersection. Section 3 describes basics of hexagonal intersection. A comparative performance analysis between both algorithms is performed in Section 4. Finally, in Section 5 some final conclusions and possible future lines of work are offered.

2   Wireless Sensor Network Localization

There is an enormous variety of positioning algorithms in the literature, depending on the exact method to estimate distances between nodes. *Time difference of arrival* (TDOA) and *received signal strength* (RSS) are the most popular methods.

Systems like Cricket [8], Calamari [15], and AHLoS [12] use the technique TDOA. They send out two signals (usually ultra sonic sound and radio frequency signals) propagating with different speeds, and measure the difference in time of arrival. If both signal propagation speeds are known, a distance can be derived from the delta of arrival. The raw difference measurements tend to yield average estimation errors of about 74% [15]. Yet, quite good accuracies can be achieved by post-processing the measured data with techniques like noise canceling, digital filtering, peak detection and calibration. While some authors report average range estimation error of 10% [15], others claim an error of about 1% at a maximum range of 9 meters [9].

While these systems yield low estimation errors they have two limitations that limit their applicability in real world deployments. The first one is their reduced coverage: They are typically able to cover 3-15 meters [11], which is only a fraction of the communication range of radio frequency transmitters. The second and much more important limitation is that they require a separate sender/receiver pair, which implies negative effects on size, cost and energy consumption.

Algorithms based on RSS use radio signal attenuation properties to model the distance between nodes as a function of the *received signal strength indicator* (RSSI). Systems that rely on the RSSI as input parameter such as [10], [2], [1] tend to be quite accurate for short ranges if extensive post-processing is employed, but are imprecise beyond a few meters [6]. At short ranges, distance estimations exhibiting error of about 10% at the maximum range of about 20 meters [16] are feasible. The uncertainty of the radio propagation imposes problems like multipath propagation, fading and shadowing effects as well as obstacles in the line-of-sight. These effects complicate the development of a consistent model. As a result systems relying exclusively on RSSI values remain inaccurate distance estimators.

Regardless of the method applied to estimate distances, the possible position of a node is represented by a circle around the beacon. Then, the intersection point of three circles, by means of a technique called *trilateration*, gives the node location (Fig. 1).

Another localization algorithms use directional antennas and triangulation methods [7]. In [13], a combination of RSS and Bayesian inference is applied.

Apart from the above techniques, there exists a collection of localization mechanisms, called *range-free*, in which it is not necessary to estimate distances to neighbor nodes. A simple option lies in that a node receiving the position of a neighbor, estimates its own one inside the coverage area of the neighbor.

Additionally, the intersection of different coverage areas contributes to reduce the potential area where the device may be located.

Obviously, a realistic approach consists in representing the coverage area by using a circle. However, the drawback of this option is that the result of intersecting two circles is not a regular geometric shape, increasing the complexity of the data structure used to represent it. One example is shown in Fig. 2.

An alternative is the *Bounding-Box* algorithm [14], in which that circular area is represented by means of a square, getting a lighter, but more inaccurate algorithm. Next, we describe the characteristics of rectangular intersection in more detail.

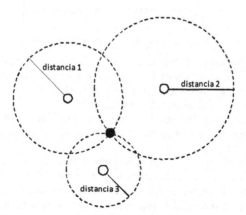

Fig. 1. Trilateration. Distances to three beacons (white nodes) allow a sensor (black node) to determine its location.

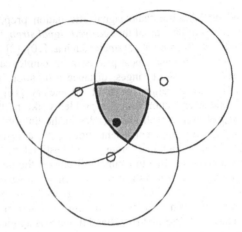

Fig. 2. Circular Intersection. Intersection of coverage areas of the three beacons allows a sensor to fence in the zone in where it is.

2.1 Rectangular Intersection

This technique is based on the idea of connectivity. If two nodes can communicate with each other, it is assumed that one is within a square centered at the other node and with a side equal to twice the radio range. The main advantage of this algorithm is that intersecting squares is a mathematically simpler operation than intersecting circles. The reason is that when intersecting squares, the result is a rectangle (Fig. 3); however when intersecting circles, the intersection is much harder to describe mathematically. Each rectangle can be represented by its upper left and lower right corners. If all neighboring nodes have their centers at coordinates (x_i, y_i), then the corners of the intersection rectangle can be defined by

$$\left[x_{UL}, y_{UL}\right] = \left[\max_{i \in 1..N}\{x_i\} - R, \max_{i \in 1..N}\{y_i\} - R\right]$$

$$\left[x_{LR}, y_{LR}\right] = \left[\min_{i \in 1..N}\{x_i\} + R, \min_{i \in 1..N}\{y_i\} + R\right]$$

The second advantage of this algorithm is that it is executed in a distributed fashion. Each node acquires positions of all its neighbors. Then, the squares centered at these neighbors are intersected yielding a final rectangle. The final position estimate is the center of the intersection rectangle.

The disadvantages of this method are the dependence on the convexity of communication regions as well as the need for high connectivity to obtain a refined and accurate estimate.

In the algorithm *N-Hop Multilateration* a phase of refinement is performed [12]. In this way, nodes without an initial estimate, due to the lack of connectivity with one beacon at least, take as reference points the estimated positions of unknown nodes in the previous phase (rectangular intersection). As a result, these nodes without connectivity estimate their position by intersecting the squares of these new reference

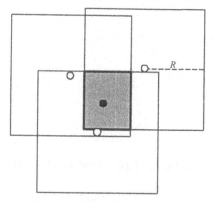

Fig. 3. Rectangular Intersection. Beacons know their exact position. The central node estimates its position from intersection of squares representing beacons range.

points, and send out new information to neighboring nodes. This process is executed in an iterative fashion until a priori established tolerance value is reached.

One further step is taken in [4]. Nodes having a location estimate do not send a pair of coordinates (x,y), but the upper left and lower right corners of the rectangle Q delimiting its location estimate. When a node receives information from its neighboring nodes, it updates its location using the next formula:

$$Q \rightarrow Q \bigcap_{k \in K} \left[x^k_{min} - S, x^k_{max} + S \right] \times \left[y^k_{min} - S, y^k_{max} + S \right]$$

where K is the set of nodes that sense that one performing the location update, S is the sensing range (R was the communication range), and $x^k_{min}, x^k_{max}, y^k_{min}$ and y^k_{max} give the coordinates of the bounding box for the position of the kth node.

3   Hexagonal Intersection

As we have seen, localization techniques based on rectangular intersection present the advantage of requiring minimum computational capacity at the nodes of the network. However, it is obvious that the error introduced in estimates is greater than using circular range areas. Given that the areas of the circle and the square are

$$A_{circle} = \pi r^2$$

$$A_{square} = (2r)^2$$

then the initial assumption of approximating a circular area by means of the square containing it, involves increasing the working area in a 27.3%.

An intermediate approach with regard to the obtained error and the complexity of the computations may consist in representing coverage areas by using hexagons. In particular, we start from the regular hexagon centered at the circular range and whose apothem is equal to the radio range (Fig. 4).

Bearing in mind that the areas of the circle and the hexagon fit the next formulas:

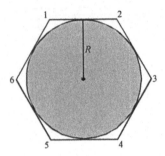

Fig. 4. Range approximation by hexagon.

$$A_{circle} = \pi r^2$$

$$A_{hexagon} = \frac{6r^2}{\sqrt{3}}$$

then, the use of this geometric shape implies an increase of a 10.2% in the localization area.

Although the result of intersecting rectangular areas in an iterative fashion is always a new rectangle, intersecting hexagonal areas will result in irregular polygons with 6, 5, 4, or even 3 sides (Fig. 5). However, these polygons have the property of that each side has a slope of 0°, 60° or 120°. Although they have not necessary six sides, from now on we call these polygons *pseudo-hexagons*.

If a rectangle is determined by the coordinates of two opposite vertices, a pseudo-hexagon can be defined by means of three vertices belonging to different sides. If we number vertices in clockwise, starting with the upper left vertex, we have considered vertices 1, 3, and 5 (Fig. 4).

Conceptually, intersection of rectangles and intersection of pseudo-hexagons are very similar. Intersection of two rectangles A and B consists in comparing each side

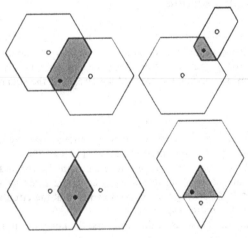

Fig. 5. Hexagonal Intersection.

of A with the corresponding side of B, and choosing those being closer to the center. After that, the computation of the four cut points of the lines defining the selected sides is performed.

The intersection of pseudo-hexagons is carried out in the same way; however, in this case, it is necessary to compare six pairs of sides (instead of four) and to obtain nine cut points.

The main difference between rectangular and hexagonal intersection is not the geometric shape itself, but the fact that the sides are or not parallel to the coordinates axes, because this implies that the computation of cut points is implicit in the case of rectangles.

In an analogous way to rectangular intersection, in order to obtain the intersection of two pseudo-hexagons, we just have to calculate vertices 1, 3, and 5 of the resulting area.

4 Performance Evaluation

To evaluate our proposal, we compare it with the improved rectangular intersection algorithm [4], described in section 2.1. Remember that in this algorithm network nodes send out the two corners determining the rectangle obtained by intersecting the rectangular areas received from other nodes.

4.1 Simulation Methodology

The simulation tool has been implemented in Adobe Flash 8. Fig. 6 shows the aspect of the user interface. The simulator gets the necessary parameters for each execution from a database, which stores simulation results as well.

We have carried out extensive numerical configurations, executing 100 simulations

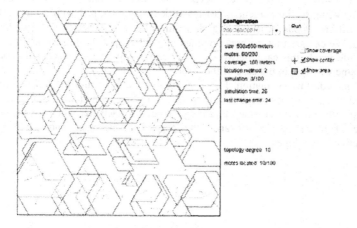

Fig. 6. Sensor Network Simulator showing localization areas and errors.

for each one and obtaining average results. In all cases, the nodes were distributed randomly in a square area of 500×500 meters; the coverage range of the motes has been set to 100 meters. Although in the rectangular intersection algorithm the communication range and sensing range are different, we have assumed the same value for these two parameters.

For the configurations, we have varied the amount of network nodes (from 100 to 300) and the amount of nodes knowing their accurate location (from 10% to 70%).

We have used two metrics to evaluate our algorithm. The first one determines the distance between the real and estimated positions, denoted as localization error. Also, we compute the size of the final area in which the node estimates its position, denoted as localization area.

4.2   Simulation Results

Fig. 7 shows the error obtained by the rectangular intersection and hexagonal intersection algorithms (identified by "R" and "H", respectively, in the legend), as a function of the percentage of network nodes which are equipped with a GPS receiver, and considering different amounts of nodes in the scenario.

In order to estimate the error for each node, we have obtained the Euclidean distance from the center of the polygon obtained –rectangle or pseudo-hexagon– by the localization algorithm to the real position of the node. For each simulation run we have computed the average error, and finally each point in the plots corresponds to the average error for a set of simulations. Fig. 7(f) shows the error rate considering both techniques. Each series in this plot has been obtained from the results shown in the previous plots.

We can observe that the error obtained when the hexagonal intersection technique is applied is about a 10% lower. In some cases, we obtain an improvement of a 30%. Additionally, this difference increases as the fraction of nodes equipped with GPS increases.

Fig. 8 shows the size of the localization area obtained by both algorithms, according again to the percentage of network nodes equipped with a GPS receiver, and for different node densities. Fig. 8(f) shows the way in which this area has been reduced. We can see that the area obtained by the hexagonal intersection algorithm is a 12% inferior. In this case, network density and the amount of beacons do not affect in the same way that in the case of the error.

To conclude this evaluation, Fig. 9 compares the error and the localization area obtained by both techniques, but in this case according to the amount of nodes in the scenario. We can observe again that the estimations are better when we increase the number of nodes in the scenario, the portion of beacons, and the hexagonal intersection is employed.

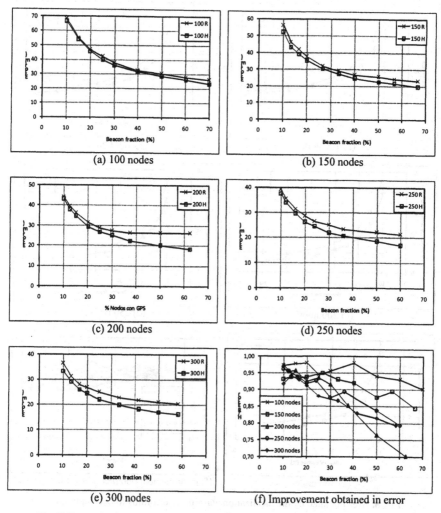

(a) 100 nodes

(b) 150 nodes

(c) 200 nodes

(d) 250 nodes

(e) 300 nodes

(f) Improvement obtained in error

Fig. 7. Error obtained by the localization algorithms, according to beacon fraction.

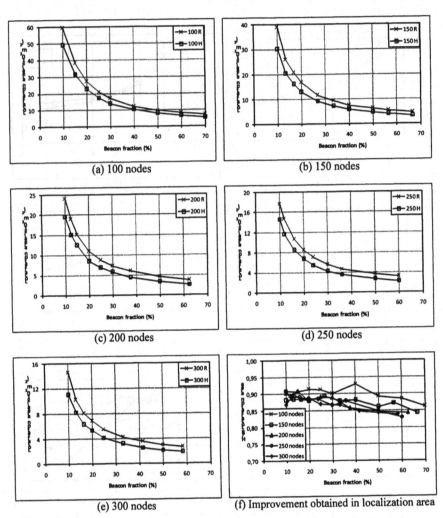

(a) 100 nodes

(b) 150 nodes

(c) 200 nodes

(d) 250 nodes

(e) 300 nodes

(f) Improvement obtained in localization area

Fig. 8. Area obtained by the localization algorithms according to beacon fraction.

5   Conclusions and future work

In this work, we have shown that the use of a new localization technique called hexagonal intersection contributes to reduce the error obtained by some traditional localization algorithms for wireless sensor networks, based on assuming square coverage areas.

As future work, we plan to implement our proposal in real new generation devices, in which additional complexity is not a problem. Also, we plan to introduce a mobile beacon in the localization system.

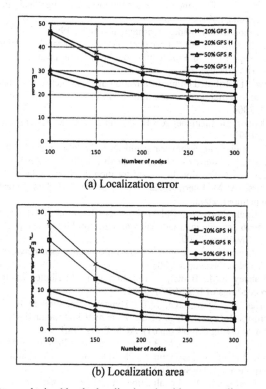

(a) Localization error

(b) Localization area

Fig. 9. Error and area obtained by the localization algorithms according to number of nodes.

6 Acknowledgements

This work has been jointly supported by the Spanish MEC and European Comission FEDER funds under grants Consolider Ingenio-2010 CSD2006-00046 and TIN2006-15516-C04-02; by JCCM under grant PBC05-007-1; and by UCLM under grant TC20070061.

References

1. Bergamo, P., Mazzini, G.: Localization in sensor networks with fading and mobility. In Proceedings of the 13th IEEE International Symposium on Personal, Indoor and Mobile Radio Communications, 2002.
2. Beutel, J.: Geolocation in a picoradio environment. Thesis, 1999.
3. Buschmann, C., et al.: Radio Propagation-Aware Distance Estimation Based on Neighborhood Comparison. In Proceedings of the 4th European conference on Wireless Sensor Networks (EWSN'07), 2007.
4. Galstyan, A., et al.: Distributed online localization in sensor networks using a moving target. In Proceedings of International Symposium of Information Processing Sensor Networks (IPSN'04), April, 2004.
5. Hoffmann-Wellenhof, B. Lichtenegger, H., Collins, J.: GPS: Theory and Practice. 4th ed. Springer-Verlag, 1997.
6. Maroti, M., et al.: Radio interferometric geolocation. In Proceedings of the 3rd International conference on Embedded networked sensor systems (SenSys'05), 2005.
7. Nasipuri, A., Kai, L.: A directionality based location discovery scheme for wireless sensor networks. In First ACM International Workshop on Wireless Sensor Networks and Applications, September, 2002.
8. Priyantha, N.B., Chakraborty, A., Balakrishnan, H.: The cricket location-support system. In Proceedings of the 6th annual international conference on Mobile computing and networking (MobiCom'00). ACM Press, 2000.
9. Sallai, J., et al.: Acoustic ranging in resource-constrained sensor networks. In International Conference on Wireless Networks, 2004.
10. Savarese, C., Rabaey, J.M., Beutel, J.: Locationing in distributed ad-hoc wireless sensor networks. In Proceedings of ICASSP'01, 2001.
11. Savvides, A., Han, C., Strivastava, M.B.: Dynamic fine-grained localization in ad-hoc networks of sensors. In Proceedings of the 7th annual international conference on Mobile computing and networking (MobiCom'01). ACM Press, 2001.
12. Savvides, A., Park, H., Srivastava, M.B.: The bits and the flops of the N-hop multilateration primitive for node localization problems. ACM International Workshop on Wireless Sensor Networks and Applications (WSNA'02), 2002.
13. Sichitiu, M.L., Ramadurai, V.: Localization of wireless sensor networks with a mobile beacon. In Proceedings of MASS, 2004.
14. Simic S.N., Sastry, S.: A distributed algorithm for localization in random wireless networks. 2002, unpublished manuscript.
15. Whitehouse, K., Culler, D.: Calibration as parameter estimation in sensor networks. In proceedings of the 1st ACM international workshop on wireless sensor networks and applications (WSNA'02). ACM Press, 2002.
16. Whitehouse, K., et al.: The effects of ranging noise on multihop localization: an empirical study. In the 4th International Conference on Information Processing in Sensor Networks (IPSN'05), 2005.

TAIL: Two-level Approach for Indoor Localization

V. Sudha Rani and S. V. Raghavan

Network Systems Laboratory
Department of Computer Science and Engineering
Indian Institute of Technology Madras, Chennai - 600 036, INDIA
{sudharani, svr}@cs.iitm.ernet.in

Abstract. Popularity of ubiquitous computing increases the importance of location-aware applications, which increases the need for finding location of the user. In this paper, we present a novel localization method for indoor environments using Wi-Fi infrastructure.

While localization using Wi-Fi is cost effective, handling the obstructions which are the main cause of signal propagation error in indoor environments is a challenging task. We address this problem in two levels, resulting in increased accuracy of localization. In the first level, we "localize" the residing area of user node in coarse granularity. Then, we use building layout to find the objects that *attenuate the signal* between the reference node and the coarse estimate of the location of user node. Using multi-wall propagation model, we apply corrections for all obstructions and *find* the location of user node. Empirical results based on experiments conducted in lab-scale, shows meter-level accuracy.

Keywords: Propagation Model, Ubiquitous Computing, Wi-Fi

1 Introduction

Use of wireless technology has become pervasive in our day-to-day life. Besides, a wide range of applications make use of *location information* to improve performance or to enhance user convenience. Localization is the process of finding out the location of a node - either by itself or by a central server. The most popular localization technology Global Positioning system (GPS) [10] is best suited for outdoors. Practical limitations such as cost, power, precision, and inaccessibility render the choice of using GPS for positioning in indoor environments infeasible. Indoor localization techniques have been proposed hitherto, differ either in *cost of deployment or in accuracy or both.* Among these, the localization techniques which use Wi-Fi technologies based on 802.11x standards obviates the need for installing special hardware. Moreover these are found almost everywhere - in buildings, in organizations, in public utility services and in common usage areas such as hospitals, airports, and restaurants. In the sequel, we propose an approach using 802.11x based equipment that provides meter-level accuracy which seems to be sufficient for most emerging applications.

Please use the following format when citing this chapter:

Sudha Rani, V., Raghavan, S. V., 2007, in IFIP International Federation for Information Processing, Volume 248, Wireless Sensor and Actor Networks, eds. L. Orozco-Barbosa, Olivares, T., Casado, R., Bermudez, A., (Boston: Springer), pp. 167-178.

The indoor localization techniques proposed thus far can be classified into two classes viz; *range-based* and *range-free* techniques. *Range-based* techniques rely on range measurements to estimate the distance to the reference nodes, using Infrared or Ultrasound or 802.11x technology and then uses trilateration or multilateration technique, to calculate the exact position of user node. On the contrary, *range-free* techniques use only the connectivity information from 802.11x reference nodes to estimate the location. *Range-based* techniques require either installation of additional hardware like Infrared or Ultrasound receivers or can lead to calibration overhead, where as *range-free* techniques require dense deployment of reference nodes to achieve appreciable accuracy.

The major source of error for localization using 802.11x infrastructure in indoor environments is the obstructions that interfere with the signal propagation from the reference node to user node. The attenuation caused is dependent on the number and type of obstructions. The accuracy of the localization method can be increased if these obstructions are identified and necessary corrections are applied to the signal strength during distance estimation. But the actual number and type of obstructions between user node and the reference node depends on the exact location of user node, which is normally unknown. To overcome this problem, a *Two-level approach* is proposed in this paper. The two levels in the proposed approach are - *Macro* and *Micro*. In *Macro level*, the *Locality[1]* of user node using *range-free* technique is computed. Then the number and type of obstructions are found using the computed *Locality* and knowledge of the building layout. In *Micro level*, corrections to the signal strength are applied using the obstructions found, which are then fed to   *range-based* technique to find the location of user node.

Depending on the applications, the accuracy required varies from coarse-grain accuracy to fine-grain accuracy. For example, finding the nearest printer in an office requires coarse-grain accuracy, where as finding the exact position of the patients in emergency conditions in a hospital requires fine-grain accuracy. The technique proposed in this work can adapt to these varying application requirements. We achieve this by adjusting the cell size; large cells for coarse-grain requirements and small cells for fine-grain requirements.

2   Related Work

As explained in previous section, work in the area of indoor localization of a node can be classified broadly into two categories, viz; *range-based* and *range-free* techniques. Cricket [1], Active Badge [2] and APS [3] fall into *range-based* category. The common principle behind the *range-based* techniques is to find the distance between user node and the reference node   using either time of flight or received signal strength (RSS) as the measure. Having found the distances to a minimum of three reference nodes, multilateration technique is applied to calculate the location of user

---

[1] Portion of a building where user is residing

node. The techniques which take time of flight as a metric, require installation of additional hardware either at user node or at the reference node. Even though RSS technique doesn't require installation of additional hardware, finding distance to the reference nodes based on received signal strength is not accurate in indoor environments due to the attenuation caused by obstructions. To overcome the difficulty of predicting signal propagation, radio map technique RADAR [4] has been proposed which works in two phases - offline and online. In offline phase the strength of the signals from the reference nodes are measured and stored in the database at all positions of the operating environment and in online phase the received signal strengths are compared with the ones stored in the database to find the location of user node. The disadvantage of this method is that tedious work is called for during offline phase and the process has to be repeated if the position of even one reference node changes which is not infrequent in real world. Besides, maintaining the database is an issue when operating in large buildings.

Proposals based on *range-free* techniques are described in [5], [6], [7] and [9]. All these techniques use *radio connectivity information* of the reference nodes in finding the location. These techniques overcome the calibration costs encountered in RSS and offline work involved in the RADAR. Besides, these techniques obviate the need for additional hardware. But, the accuracy of *range-free* techniques is less compared to RSS-based techniques, as it uses only connectivity information. Appreciable accuracy of *range-free* localization techniques can be achieved only when the density of reference nodes is more or when reference nodes are distributed uniformly, or when the number of neighboring user nodes is more [9]. Satisfying these conditions at all times cannot normally be guaranteed in practical situations.

The work reported in this paper overcomes all the above limitations and still gives appreciable accuracy. We demonstrate through experiments that the error-in-estimation in the proposed technique is less than two meters.

3   Wireless Channel Characteristics

The basic property of the radio propagation is that, the received signal strength value decreases as the distance between transmitter and receiver increases which makes it a suitable metric for relative-distance measurement. The distance to signal-strength relation is captured in signal propagation model. Most widely used propagation model for indoor environments is *log distance path loss model* [8], which is given as:

$$RSS(d) = Pt - PL(d_0) - 10\eta \log_{10}(d/d_0) .$$ (1)

In this equation *RSS(d)* is the strength of the signal received from the reference node at distance of d meters, P_t is the transmitted power, $PL(d_0)$ is the path loss for a reference distance of d_0 [8], and η is the path loss exponent. While usage of this propagation model to predict the signal strength is simple, it doesn't take into account the operating environment which impacts radio propagation. Therefore, propagation model captured in equation (1) is best suited for obstruction free environments but not for indoor environments where obstructions are abound.

To consider the effect of obstructions on the signal strength, multi-wall propagation model [8] is used in the proposed work. The building layout data is used to identify the obstructions that are present in the path between transmitter and receiver. This in turn can be used to make necessary corrections to the received signal strength, based on the number and type of obstructions present between the transmitter and receiver. The generalized propagation model can then be given as:

$$RSS(d) = P_t - PL(d_0) - TL(d) \ . \tag{2}$$

where $TL\ (d)$ represents the total transmission loss for a particular distance d which is composed of two components: the path transmission loss $(PL\ (d))$ and the loss due to obstructions (OL). Hence the total transmission loss can be given as:

$$TL(d) = PL(d) + OL \ . \tag{3}$$

Obstruction Loss (OL) depends on the number of obstructions between the transmitter and the receiver and is sum of attenuations which is given as:

$$OL = \sum_{i=1}^{n} m_i x_i \ . \tag{4}$$

where m_i is the number of obstructions of a particular type i and x_i is the attenuation caused by it. OL drops down to zero when line of sight conditions exist between transmitter and receiver and increases with increase in the number of obstructions. Therefore multi wall propagation model is given as:

$$RSS(d) = P - PL(d_0) - 10\eta \log_{10}(d/d_0) - \sum_{i=1}^{n} m_i x_i \ . \tag{5}$$

The parameter x_i for each type of obstruction was determined empirically from a series of experiments conducted under controlled conditions. Next, we will explain the experimental setup followed by the results obtained.

We conducted two experiments, one to show that log normal propagation model holds good when there is line of sight between sender and receiver and the second one to understand the effect of obstruction on the signal from sender to receiver. The experiments were called Line Of sight (LOS) Propagation Model experiment and Non-LOS propagation Model experiment respectively.

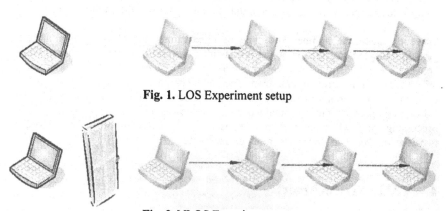

Fig. 1. LOS Experiment setup

Fig. 2. NLOS Experiment setup

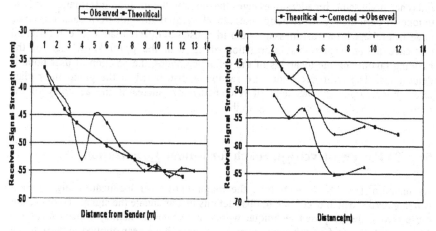

Fig. 3. LOS Propagation Model   **Fig. 4.** NLOS Propagation Model

The experimental setup for these two experiments is shown in Fig. 1 and Fig. 2 respectively. Two laptops were taken, of which one was the reference node and the other was user node. NetStumbler software was installed in user node which captured the signal strength of the broadcast messages sent by the reference node. The signal strength readings were measured at various distances from the reference node. In LOS experimental setup, Line of Sight condition exists between the reference node and user node, where as in NLOS setup, an obstruction was kept between the reference node and user node to find the attenuation loss of this obstruction.

We observe from Fig. 3 that, in LOS experiment the theoretical and observed values correlate each other. In the NLOS case, we observe (Fig. 4) that the observed values shifted and are aligned with theoretical values after the correction was applied. The hump in these figures can be attributed to the random noise in the environment.

However, since the number and type of obstructions cannot be known apriori, globally accepted propagation model is difficult to obtain. The actual number of walls between user node and the reference node depends on the exact location of user node, information that is normally unknown. The naive approach to estimate the position is to find the *signal strength vector* at various points using multi wall propagation model [8] and compare it with the obtained one. The downside of this approach is large processing time, when the building size is more. To overcome this, we propose a *two-level approach* to find the location of user node.

4   TAIL: System Model

In this work, we consider the localization system consisting of a set of 802.11b reference nodes and a set of user nodes located on a 2-dimensional floor-plan. Each reference node sends broadcast messages that contain the location information and the unique Id periodically. User nodes measure the signal strength from each reference node from these broadcast messages and Id is used to differentiate between different reference nodes. Using this information along with the priori knowledge of the building layout and the locations of the reference nodes, the position of user node is determined. The reference nodes are randomly distributed in the given floor-plan. Even though optimal placement of reference nodes increases the accuracy of the proposed method, it is not examined in this work.

5   TAIL: Two-level Approach for Indoor Localization

As mentioned earlier in section 1, the major hurdle for localization algorithm is *ranging errors* which occur due to obstructions that attenuate the signal. To overcome these errors, the proposed technique works in two levels namely *Macro level* and *Micro level*. *Range-free* and *range-based* techniques have been applied in these levels respectively. *Macro level* estimates the *Locality* of user node, the size of which depends on the number of reference nodes in range. *Micro level* takes this *Locality* information as input, computes number and type of obstructions, and finally uses corrected multi-wall propagation model (equation 5) to compute the location of user node. The following subsections explain each level in detail.

The proposed technique requires the storage of only the building layout and the attenuation loss of the obstructions present in the building, making it suitable to implement either in a centralized manner - where a central server finds the location of all user nodes, or in a distributed manner - where user node finds its own location. It is assumed that network level connectivity exists between user node and central server and the underlying communication technology is 802.11x. The choice between centralized and distributed approaches can be made depending on the requirements of the application, power constraints of user nodes and privacy issues involved. Distributed implementation requires, every user node to be provided with building layout and the attenuation values of the obstructions. In a centralized implementation, central server stores this information and either user node transmits the received signal strength readings to the server, or the reference nodes measures the signal strength of the messages from user node and sends them to the central server.

5.1   Macro Level

Macro level applies *range-free* technique to estimate the *Locality* of user node. *Range-free* technique uses the overlapping coverage region of reference nodes to estimate the required area [7]. The reference nodes having IDs - ID_1 to ID_n are situated at positions (X_1, Y_1) to (X_n, Y_n) with transmission ranges R_1 to R_n respectively. These reference nodes broadcast messages periodically which contain their ID and location coordinates. User node listens for these broadcast messages for a period of time and lists the reference nodes which it is able to hear. User node can receive the broadcast messages of the reference node if and only if it is in the coverage region of the reference node, which is the circle drawn with the reference node location as the center and its transmission range as the radius. If user node is able to hear from more than one reference node, then the overlapping coverage region of all these reference nodes correspond to the *Locality* of user node. Here it has to be observed that the signal strength of the messages from the reference node is not being taken into consideration, so the accuracy of the returned *Locality* will not be affected by the attenuation error caused by the obstructions.

This is shown pictorially in Fig. 5, where A1, A2 and A3 are the reference nodes and user node is receiving broadcast messages from all the three nodes. The *Locality* of user node is shown as shaded region in Fig. 5. Increase in the number of reference nodes in its range, results in finer granularity of the region. In the following paragraph, we present the Cell Table approach which returns the *Locality* as the set of cells.

The whole operating environment is divided into cells as shown in the Fig. 6, where each square represents one cell. Each cell is assigned an integer called cell rank which is set to zero initially for all the cells. This cell rank represents the number of reference nodes in its range. Each cell is checked for the validation of In-Range property detailed below to decide if the cell is included in the transmission range of the reference node. If the cell satisfies this property, the rank of the cell is incremented by one otherwise the value remains unchanged. This process is repeated for all the reference nodes from which user node has received the broadcast messages.

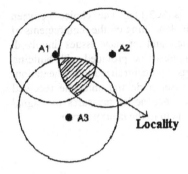

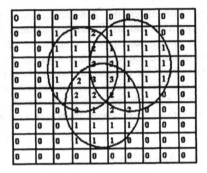

Fig. 5. Intersection Approach   **Fig. 6.** Cell Table

The In-Range property checks whether the cell is in the coverage region of the corresponding reference node or not, which can be formulated as below:

$$P : (x_c - x_i)^2 + (y_c - y_i)^2 <= R_i^2 . \tag{6}$$

In this equation, (x_c, y_c) are the coordinates of the center of cell, (x_i, y_i) are the coordinates of one of the reference nodes from which the node has received broadcast messages and R_i is its transmission range. The property is satisfied when the distance between the center of cell and the reference node is less than its transmission range.

The overlapping region i.e. the *Locality* of user node is defined as the set of cells which are having the highest rank in the cell table. In the Fig. 6, user node is in the range of three reference nodes, so the *Locality* corresponds to the set of these cells whose rank is three. The pseudo code for Cell Ranking is as follows:

For each cell C of the building
   {
        For each reference node ID_i user node is hearing
           {
               if $((x_c\text{-}x_i)^2 + (y_c\text{-}y_i)^2 - R_i^2 <= 0)$
                  then increment cell rank by 1;
           }
   }

5.2   Micro Level

Micro level estimates the cell in which user node is residing from the set of cells obtained in *Macro level*. As said previously, most of the error in indoor environments is due to the attenuation caused by obstructions between the transmitter and the receiver. If the number of obstructions between the reference node and user node were known, then the distance to the reference node can be determined using the multi-wall propagation model given in equation (5) by applying the correction as explained in section 3. However, the actual number of obstructions between the reference node and user node depends on the exact location of user node and is thus unknown. To circumvent this problem, for all the cells obtained in the *Macro level*, *signal strength vector* is estimated by finding the obstructions and using equation (5), which is then compared with the received *signal strength vector*. The location of user node is defined as the center of the cell whose estimated *signal strength vector* best matches with the received *signal strength vector*. This is explained in detail with the help of an example in the following paragraph.

Consider the layout plan shown in Fig. 7, where reference nodes are shown as A1, A2 and A3 and square C represents one of the cells obtained from the *Macro level*. The obstructions between cell C and each of the reference nodes can be determined as the positions of cell, reference nodes and layout plan is known. For example, in the Fig. 7 shown, the number of obstructions from the cell center to the reference nodes A1, A2 and A3 are 1, 3 and 3 respectively. Taking this information into consideration *signal strength vector* is estimated using equation (5).

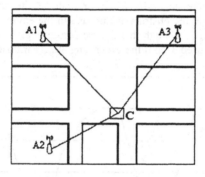

Fig. 7. Building Layout

This *signal strength vector* is estimated for each cell obtained from the *Macro level*. It is of the form $S_i = (s_{i1}, s_{i2}, s_{i3} \ldots s_{ik})$, for all $i=1$ to N, where N is number of cells obtained in *Macro level* and k is the number of reference nodes covering the obtained cells and s_{ij} denotes the signal strength at cell i from the jth reference node. The received *signal strength vector* is represented as $R = (r_1, r_2, r_3 \ldots r_k)$. The Euclidean distance given in equation (7) is calculated between the received *signal strength vector* and the estimated *signal strength vector* for all the cells.

$$D(S_i, R) = \sqrt{\sum_{j=1}^{k} (S_{ij} - r_j)^2} \ . \tag{7}$$

$$CellNumber = \min_i(D(S_i, R)) \ . \tag{8}$$

The *CellNumber* with minimum Euclidean distance is found as shown in equation (8), whose center is taken as the location of user node.

Extension of the proposed method to 3-dimensional environment is straight forward by taking z-coordinate in equation (6) and taking attenuation loss due to floors between transmitter and receiver into account in multi wall propagation model given in equation (5).

6   TAIL: Experimental Results

We conducted a series of experiments under controlled conditions to evaluate the performance of TAIL. All the experiments are conducted in two-dimensions by keeping all the reference nodes and user node at the same height. Initially the attenuation loss of all the obstructions present in the environment were calculated as explained in section 3. Though this is time consuming, it is a one-time process and these values can be used even if the underlying 802.11 infrastructure changes.

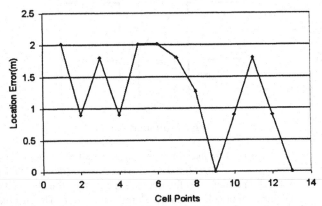

Fig. 8. Location Estimation Error for various points

The considered infrastructure contains four reference nodes and one user node. The reference nodes were static. Netstumbler software package was installed in user node to capture the signal strength of the messages sent by the reference nodes. The building was divided into cells of size 0.9 x 0.9 m. User moved along various cells and measured both the actual position and signal strength readings from the reference nodes. The location of user node was estimated with these *signal strength vectors* using the proposed *two-level* approach. The location estimation error was calculated as the distance between the actual location and estimated location. Let (x, y) represent the actual location and (x_{est}, y_{est}) represent the estimated location. The error is given by

$$Error = \sqrt{(x_{est} - x)^2 + (y_{est} - y)^2}$$ (9)

The graph in Fig. 8 shows the location error for various points. The x-axis shows various points in the building and the y-axis shows error in meters. It can be observed from Fig. 8 that the error of the proposed technique is less than 2 meter for all the cells.

7.   TAIL: Simulation Results

We have done simulation in Matlab to see the effect of number of reference nodes and size of cell on location accuracy. Size of the building was taken as 60x30 m^2 and parameter values of the multi-wall propagation model used are given in Table 1. A random noise with $\sigma^2 = 4$ dB was added to the model and positions of the reference nodes and user nodes were selected randomly inside the building.

Table 1. Parameter Values for Propagation Model

d_o	1 meter
P_t-PL(d_0)	-38 dBm
η	2.5
R	30 meters
x_i	4 dB

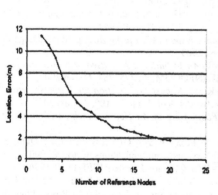

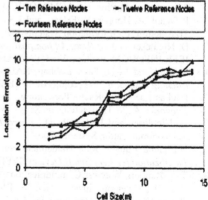

Fig. 9. Location Error against number of Reference Nodes

Fig. 10. Location Error against Cell size under varying number of Reference Nodes

Fig. 9 shows that, the location error decreases with the increase in number of reference nodes. As the total number of refernce nodes increases, the number of reference nodes that are in the range of user node also increases which results in the decrease in location error. Fig. 10 shows the effect of cell size on location accuracy when the number of reference nodes are ten, twelve and fourteen. It shows that, the smaller the cell size, the better the accuracy of the system irrespective of the number of reference nodes. This decrease in location error is attributed to the increase in the number of points where *signal strength vectors* are estimated and compared with the *received signal strength vectors*. Therefore large cell size has to be chosen for coarse-grain requirement and small cell size for fine-grain requirements.

8.   Conclusions

In this paper we have proposed a technique called TAIL which works in two levels using both *range-free* and *range-based* techniques in a sequence. TAIL uses the *range-free* technique to find the *Locality* of user node and then use multi-wall propagation model inside this *Locality* to find out the exact position. TAIL provides improved accuracy because we are using multi-wall propagation model which corrects the attenuation caused by obstructions in the building. We have shown through experiments that the error obtained is less than two meters, which is appreciable; and processing required is less than the methods present hitherto, as the exact location is estimated from the *Locality* identified by the *Macro-level* part of the TAIL as opposed to considering the whole operating environment.

References

1.   Nissanka B. Priyantha, Anit Chakraborty and Hari Balakrishnan, *The Cricket Location-Support system*, Proc. 6th ACM MOBICOM, August 2000.
2.   R. Want, A. Hopper, V. Falcao, and J. Gibbons, *The active badge location system*, ACM Transactions on Information Systems, vol. 10, pp. 91--102, Jan. 1992
3.   D. Niculescu and B. Nath, *Ad hoc positioning system (APS) using AoA*, In Proceedings of IEEE INFOCOM 2003, pp 1734-1743, April 2003.
4.   P. Bahl and V. N. Padmanabhan, RADAR: *An in-building RF-based user location and tracking system*, In proceedings of the IEEE INFOCOM, pp 775–784, March 2000.
5.   Loukas Lazos and Radha Poovendran, SeRLoc: *Robust Localization for Wireless Sensor Networks*, ACM Transactions on Sensor Networks, Vol. 1, pp 73–100, August 2005.
6.   D. Niculescu and Badrinath, *DV Based Positioning in Ad Hoc Networks*, In Journal of Telecommunication Systems, 2003.
7.   Nirupama Bulusu, John Heidemann, and Deborah Estrin, *GPS-less Low-Cost Outdoor Localization for Very Small Devices*, IEEE Personal Communications, October 2000
8.   T. S. Rapport, Wireless Communications – Principles and Practice, IEEE Press 1996
9.   Tian He, Chengdu Huang, Brian M. Blum, John A. Stankovic and Tarek Abdelzaher, *Range-Free Localization Schemes for Large Scale Sensor Networks*, In Proceedings of MobiCom '03, September 2003.
10. P. Enge and P. Misra, *Special Issue on GPS: The Global Positioning Systems*, Proceedings of the IEEE, January 1999.

Low Overhead Assignment of Symbolic Coordinates in Sensor Networks

Matthias Gauger[1,2], Pedro José Marrón[2], Daniel Kauker[1], and Kurt Rothermel[1]

[1] IPVS, Universität Stuttgart, Germany
{gauger, kaukerdl, rothermel}@ipvs.uni-stuttgart.de
[2] Universität Bonn, Germany
{gauger, pjmarron}@cs.uni-bonn.de

Abstract. Approximate information on the location of nodes in a sensor network is essential to many types of sensor network applications and algorithms. In many cases, using symbolic coordinates is an attractive alternative to the use of geographic coordinates due to lower costs and lower requirements on the available location information during coordinate assignment. In this paper, we investigate different possible methods of assigning symbolic coordinates to sensor nodes. We present a method based on broadcasting coordinate messaging and filtering using sensor events. We show in the evaluation that this method allows a reliable assignment of symbolic coordinates while only generating a low overhead.

1 Introduction

Information on the context of nodes in a sensor network is essential to many types of sensor network applications. Examples of such context knowledge are the positions of nodes, their neighborhood or the external conditions of a node's surroundings. Among the different types of context, location information plays an especially important role in sensor networks as it is required if sensor readings are to be associated with the area they were recorded in.

Acquiring position information of sensor nodes in the form of geographic coordinates with an acceptable precision is a very difficult and often costly operation. This is especially true for indoor scenarios where localization technologies like GPS receivers do not work well. One possible alternative to determining geographic coordinates is to assign symbolic coordinates to nodes. Instead of describing positions in the form of a coordinate tuple, a symbolic coordinate represents areas of different shapes and sizes in the form of an abstract symbol. All sensor nodes in an area have the same symbolic coordinate. Examples of such symbolic coordinates are room numbers in a building or street addresses.

In different types of sensor network applications it is possible to use symbolic coordinates instead of geographic coordinates. One exemplary field is the retrieval of sensor data from specific areas of a sensor network. In many cases, symbolic coordinates directly represent the semantics of a location, for example, when a symbolic coordinate is associated with each room of a building. This

Please use the following format when citing this chapter:

Gauger, M., Marrón, P. J., Kauker D., Rothermel, K., 2007, in IFIP International Federation for Information Processing, Volume 248, Wireless Sensor and Actor Networks, eds. L. Orozco-Barbosa, Olivares, T., Casado, R., Bermudez, A., (Boston: Springer), pp. 179-190.

allows to implement data retrieval operations very easily without having to map from coordinates to areas first. Another possible application of symbolic coordinates is using them for cost-effective many-to-many routing of messages in sensor networks. We specify symbolic source routes from the sender node to the destination area and later translate these routes on the node level into specific routing decisions. The applications of the methods presented in this paper do not have to be limited to the assignment of symbolic coordinates. It is also possible to use them for node clustering (i.e., all nodes in a room form a cluster). In such room-level clusters the nodes typically generate related information which can be aggregated with less information loss than in arbitrary clusters.

We argue that assigning symbolic coordinates in sensor networks is usually much easier than assigning geographic coordinates and is possible with reasonable effort. In this paper, we discuss different approaches to this assignment for indoor scenarios and present one solution that achieves a very low error rate while only generating a small overhead and requiring no prior knowledge on the sensor network topology. The basic idea is to let an administrator broadcast symbolic coordinates in the different rooms a sensor network is deployed in and let the nodes use sensor information to filter out broadcast messages wrongly received from neighboring rooms.

The rest of this paper is organized as follows. The following section briefly reviews important related work. We present our three approaches to the assignment of symbolic coordinates in Section 3. In Section 4 we describe relevant details of our implementation before Section 5 provides an in-depth evaluation of our approaches. Section 6 concludes the paper and discusses some future work.

2 Related Work

Our work is related to the large set of approaches in the area of node localization which is one of the fundamental research problems in wireless sensor networks. Most localization approaches have in common that they require a set of anchor nodes with known positions. Then different techniques and algorithms are used to determine the positions of the other nodes (see for example [1]). One important factor is how distances between nodes are determined. There are range-free solutions that only consider the hop-count, distance estimations based on the received signal strength and solutions requiring special ranging devices (e.g., [2]). Elnahrawy, Li and Martin [3] discuss the fundamental limits of localization techniques based on signal strength when used in indoor scenarios.

The Spotlight localization system [4] is particularly related to our approach in that it also uses sensor events for the localization of nodes. A helicopter (the Spotlight device) which knows its own position flies over the sensor network and generates light events at certain points of time. The sensor nodes report when they detect events back to the helicopter, which is then able to compute the geographic coordinates of the nodes. However, as the authors aim to calculate geographic coordinates the required calculations are rather complex and a precise time synchronization of nodes is required.

Similar to our assignment of symbolic coordinates, Corke, Peterson and Rus [5] assign geographic coordinates to sensor nodes using radio communication. In their scenario, a robot helicopter equipped with a GPS receiver flies over the network area and periodically broadcasts beacon messages containing its current geographic coordinate. The sensor nodes on the ground typically receive multiple such beacons and need to estimate their position based on this information. The authors propose different methods for this calculation, including taking the mean of the positions or the signal strength weighted mean of the positions. A constraint-based method performed best in their experiments.

There has been active work on symbolic coordinates, their applications and underlying location models mostly in the area of pervasive computing (e.g., [6], [7]). Becker and Dürr [8] give a comprehensive overview of different geometric and symbolic location models from the perspective of pervasive computing and compare their suitability for different types of queries. In our work we assume that such a symbolic location model is available so that an administrator can assign symbolic coordinates to individual rooms.

3   Assignment of Symbolic Coordinates

In this section we introduce our three different approaches to assign symbolic coordinates in sensor networks. We start by describing the properties of our target scenarios in more detail.

Our target system consists of a set of sensor nodes that are distributed to different rooms in a building. Each room is uniquely identified by a symbolic coordinate, which can, for example, correspond to the respective room number. In the following discussion, we assume for simplicity that a symbolic coordinate can be represented by an integer value.

We assume that each sensor node is preprogammed with a node identifier (node ID) and that this node ID is unique in the sensor network. However, we do not assume that there is any relation between this node ID and the location of the node in the network. In the beginning, the sensor nodes do not have any information about the symbolic coordinate of the room they are located in.

After the deployment of the sensor nodes there is one person – which we call administrator – responsible for assigning the correct symbolic coordinates to the nodes in the network. The administrator has a mobile client device that is able to directly communicate with the nodes of the sensor network. The client device can be used to send so-called coordinate messages to the nodes.

3.1   Individual Assignment of Symbolic Coordinates

The most basic way of assigning symbolic coordinates to nodes is to assign the respective coordinate to each node individually. In some scenarios it might be possible to directly encode the symbolic coordinate in the program code of the node like it is typically done with the node ID. However, this already limits the flexibility for the placement of nodes. Moreover, we generally expect the sensor nodes to be delivered to users preprogrammed with an application software when the application field lies outside of typical research settings.

To assign symbolic coordinates to sensor nodes after the deployment of the network the administrator needs to send the correct symbolic coordinate in a coordinate message to each individual node using the node's wireless interface. Upon receiving such a message the node stores the new symbolic coordinate and uses it from this point on.

The clear advantage of an individual assignment of symbolic coordinates to nodes is that it avoids ambiguity: The administrator has complete control over which nodes receive which symbolic coordinate and is able to ensure the correctness of the assignment process. However, there are also a number of clear disadvantages. The assignment of coordinates requires individual communication with each sensor node in the network. Consequently, the required time and effort (and also the message complexity) grow with the number of nodes in the network. More important, it is necessary that the administrator has an up-to-date knowledge on the distribution of nodes to the individual rooms in the building.

3.2   Assignment of Symbolic Coordinates by Broadcast

Our second approach, the assignment of symbolic coordinates by broadcast, aims to address the disadvantages of having to separately assign a symbolic coordinate to each individual node in the network. Instead, the goal is to distribute the coordinate information to all nodes in the area in one step.

The network administrator needs to visit the different rooms covered by nodes of the network and has to send out a message containing the current symbolic coordinate information in each room. The message is sent by broadcast so that all nodes in the one-hop neighborhood of the client device receive the information. Upon receiving such a message, a sensor node overwrites its symbolic coordinate information with the newly received data.

One advantage of sending out coordinate information by broadcast is the lower overhead both for the administrator and in terms of messages as the configuration has to be done only once per room instead of once per sensor node. Moreover, no information is required on the position of individual sensor nodes since the coordinate information does not have to be addressed to specific nodes. The only information that must be available is the symbolic coordinate of the room the broadcast message is sent in.

It is typically desirable that sensor networks are connected across area boundaries like walls between rooms to provide for communication between different network parts. Therefore, the main challenge in the assignment of symbolic coordinates by broadcast is nodes that receive coordinate messages from neighboring rooms. Depending on the sequence of messages sent, these nodes might overwrite the correct coordinate information with data belonging to a neighboring room.

The basic approach to address this challenge is to control the signal strength of the messages sent by the client device. Ideally, the strength of the signal should still allow the message to reach all nodes inside the current room and none of the nodes in neighboring rooms. However, it is difficult to find this balance especially if two nodes in different rooms are located very close to each other and the attenuation of the signal by the wall between them is small.

We have developed two extensions for the assignment of symbolic coordinates by broadcast. The first extension allows sensor nodes to store and manage multiple symbolic coordinates at the same time so that different coordinate messages do not overwrite each other's information. If a node has received multiple symbolic coordinates, then it lies in the border area of the rooms represented by these coordinates.

The second extension analyzes the signal strength of the different coordinate messages at the receiver, called Received Signal Strength Indication (RSSI) value, and uses this information to assign a coordinate to the node (i.e., the coordinate from the message with the highest RSSI value). However, this only works reliably if the signal attenuation of the walls between rooms is significant. This can be problematic because RSSI – despite its strong limitations in indoor scenarios [3] – is still rather an indicator for geographic distances among nodes than for the separation of nodes to different areas.

3.3 Assignment of Symbolic Coordinates by Assisted Broadcast

Both approaches to the assignment of symbolic coordinates presented so far have disadvantages. The first approach is quite intricate and requires detailed knowledge on the nodes' positions in the network whereas the second approach is susceptible to errors due to the propagation of coordinate messages across room boundaries. In the following we present a third approach which avoids these problems.

Like in the previous approach, the client device broadcasts coordinate messages. It is again the task of the administrator to set the symbolic coordinate sent out by the client in a way so that it corresponds to the current location of the administrator. The important difference lies in the handling of a received coordinate message by the sensor nodes. Instead of directly assigning a symbolic coordinate received in a coordinate message, a node first checks whether the new coordinate is confirmed by a sensor stimulus following the message. For example, the node checks whether the light level changes significantly after the receiving of the message.

Directly after sending out a coordinate message (within a specified time interval of a few seconds) the administrator has to trigger an event that can be detected by one of the sensors of the nodes. The goal is to distinguish nodes that are located in the same room as the administrator and that receive the coordinate message and detect the following sensor event from nodes in neighboring areas that only receive the coordinate message but are not affected by the sensor event triggered by the administrator.

Fig. 1 shows the pseudo code for checking whether a received symbolic coordinate should be assigned. Directly after having received a new coordinate message, the node records its current sensor value and starts a timer. When the timer fires, it records its sensor value again and calculates the absolute difference

```
int ownSymbolicCoordinate;
int candidateSymbolicCoordinate;

event receivedCoordinateMessage(int newSymbolicCoordinate) {
  candidateSymbolicCoordinate = newSymbolicCoordinate;
  initialSensorValue = getSensorValue();
  startTimer(eventDetectionTimerLength);
}

event timerFired() {
  int finalSensorValue = getSensorValue();
  if (abs(currentSensorValue - finalSensorValue) > eventThreshold) {
    ownSymbolicCoordinate = candidateSymbolicCoordinate;
  }
}
```

Fig. 1. Checking a new symbolic coordinate for applicability

of the two sensor values. If this value is above a specified event threshold, the sensor node detects an event and assigns the new symbolic coordinate.

The method described above relies on two important assumptions. First, it assumes that it is possible for the administrator to change the external conditions in a way that allows all sensor nodes in the room to detect these changes as an event. Second, it also assumes that similar changes to the external conditions in the neighboring rooms are unlikely to happen at the same time without explicit intervention by an external party.

The main advantage of the third approach is that the additional sensor stimulus triggered by the administrator prevents ambiguities in the assignment of the symbolic coordinates. This allows to assign the coordinate to a specific set of nodes without having to address each of the nodes in this set individually. However, this comes at the cost of the additional effort required for generating the external sensor stimulus. Moreover, it only works for actual sensor nodes that possess the sensor chip required for detecting the event. The first two approaches also work on other devices that are part of the sensor network, for example gateway nodes without any sensing functionality.

4   Implementation

We have implemented the three approaches described above for Tmote Sky sensor nodes based on the TinyOS 2.0 operating system. The Tmote Sky sensor nodes provide three different sensor chips: The SHT1x sensor chip from Sensirion for measuring temperature and humidity as well as the two light sensors S1087 and the S1087-01 from Hamamatsu. The first sensor captures the photosynthetically active radiation (PAR), the other the total solar radiation (TSR). For the evaluation of our third approach we rely on the TSR light sensor as it is better suited for capturing artificial light than the PAR light sensor.

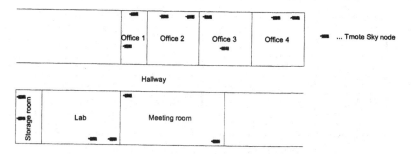

Fig. 2. Floor plan of the experiment area

The client application is implemented in C++ on the Qtopia application platform. It runs on Linux PDAs from Sharp (Sharp Zaurus SL-3200) that communicate with the sensor network using a Tmote Sky sensor node connected to the PDA over USB as a bridge node. In the future, we expect the availability of devices that are able to directly communicate with sensor nodes using a communication standard like IEEE 802.15.4. The client application supports all three approaches presented in this paper. Thus, symbolic coordinates can be sent to individual nodes or broadcasted with or without advertising a following sensor stimulus. The event detection threshold and the event detection timeout can be freely configured.

The output power of the CC2420 radio chips used by the Tmote Sky sensor nodes is programmable. The possible range of values starts with a minimum output power of -25 dBm and goes up to a maximum output power of 0 dBm. The client application allows to select any of the valid output power values.

5 Evaluation

In this section we present an evaluation of our different approaches to the assignment of symbolic coordinates in sensor networks. However, we do not discuss the first approach in detail since the manual assignment of coordinates to individual nodes should work in all cases as long as no packets are lost due to the unreliability of the wireless channel.

For evaluating our approach we deployed 14 sensor nodes in 7 different rooms of the computer science building at the Universität Stuttgart. Fig. 2 shows the floor plan and outlines the location of the nodes in the rooms. Note that we tried to create a somewhat irregular distribution of nodes with different distances among nodes in different rooms.

In a first set of experiments we investigated how reliable the assignment of symbolic coordinates using broadcast works. For this purpose, we sent out coordinate messages in each room with different output power levels and collected information on which nodes received the coordinate message with which RSSI value. We repeated each experiment five times for each signal strength level of the client device.

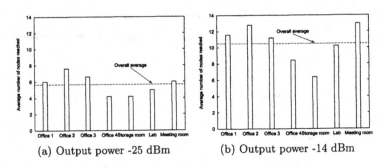

(a) Output power -25 dBm   (b) Output power -14 dBm

Fig. 3. Average number of nodes reached by coordinate message broadcasts

As a first result, these experiments showed that limiting the dissemination of coordinate messages to a single room is hardly possible even when using the minimum transmission power of the Tmote sensor nodes. To illustrate this, Fig. 3 (a) shows the average number of nodes reached from each room with the minimum output power of approximately -25 dBm and Fig. 3 (b) shows the same analysis for an output power of approximately -14 dBm. Even for the minimum output power, the overall average of 5.66 nodes reached from each area is much higher than the two nodes actually located in each area. On average, each node received coordinate messages from 2.83 different areas (5.26 for an output power of -14 dBm) with a maximum of 5 (7 for an output power of -14 dBm).

To deal with multiple coordinate messages received from different rooms we proposed to consider the RSSI values of the received messages and assign the symbolic coordinate from the message with the highest RSSI value. Using this extension results in a promising performance of the symbolic coordinate assignment using broadcast: Table 1 shows the maximum and the minimum number of nodes with correctly assigned symbolic coordinates for the experiments with signal output powers of -25 dBm and -14 dBm as well as the average percentage of nodes assigned correctly.

Table 1. Average performance for different sender signal strengths

	Output power -25 dBm	Output power -14 dBm
Max. # of nodes assigned correctly	13	13
Min. # of nodes assigned correctly	11	10
Avg. % of nodes assigned correctly	88.57	77.14

Coordinate assignments with the smaller transmission power level clearly outperform the assignment with a higher transmission power level and only assign the wrong coordinate to between one and three nodes. An explanation for this can be found looking at the RSSI values: On average, the RSSI values of messages sent in the same room are only 6% (8% for -25 dBm) larger than

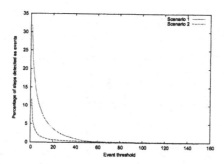

Fig. 4. Percentage of events detected for different event thresholds

the RSSI values of messages sent from different rooms. While the ratio between inside and outside RSSI values is a little higher for -14 dBm than for -25 dBm this cannot compensate the much higher number of coordinate messages each node receives. Due to the inherent variations of the RSSI values of received messages a larger number of coordinate messages received from neighboring rooms increases the probability that the RSSI value of one of these messages is larger than the RSSI value of the message received from the own room.

Overall, while the results of our second approach together with an analysis of the RSSI values produced good results in our experiments our analysis also made clear that RSSI is a fragile criterion that is not able to produce 100% reliable results.

Based on the results of the described experiments we next investigated our third approach – the assignment of symbolic coordinates by assisted broadcast.

First, we wanted to investigate how often TSR light events typically occur when they are not explicitly triggered by the user. For this purpose, we collected sensor data in two indoor scenarios. We distributed 12 nodes to 4 different rooms in both scenarios and collected the value of the TSR light sensor every 10 seconds over multiple days with each sensor node. In the analysis, we evaluated how often an event is detected for two consecutive measurements when varying the event detection threshold. Fig. 4 shows the result of this analysis for both scenarios.

The results show that for very small event thresholds a considerable percentage of measurement pairs triggers events. However, with an event threshold of 10 only in 8.1% (1.4%) and with an event threshold of 20 only in 3.7% (0.7%) of the cases an event has been detected. This indicates a quite small probability of unintended events occuring during the coordinate assignment using these thresholds especially since the event detection time period (10 seconds) was selected quite large in this case. Since changes to the level of illumination of the sensor nodes during the daytime are the main sources of events, the considerable differences between scenario 1 and scenario 2 can be explained by the fact that more nodes in scenario 1 were exposed to sunlight than in scenario 2.

Besides unintentionally detected events a second potential issue is how to trigger light events when the room is already brightly lit by sunlight coming

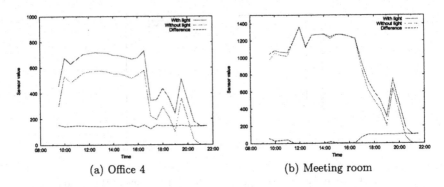

(a) Office 4   (b) Meeting room

Fig. 5. Light sensor data with and without room light

through the window. To investigate how big of a problem this is we recorded TSR light sensor values in two of our rooms for one day. One recording was done in office 4 which has its windows to the north. To explore more extreme conditions we placed the sensor for the second recording directly behind one of the south-bound windows of the meeting room. Values were recorded every 30 minutes on a sunny day both with the light turned on and with the light turned off. Fig. 5 shows the resulting graphs for both rooms.

The difference between the sensor values recorded with and without light is relatively stable over the day in office 4 (Fig. 5 (a)). With difference values in the range of 140, a reliable detection of light events is possible irrespective of the time of day the coordinate assignment is done. The situation is different for the sensor at the window of the meeting room (Fig. 5 (b)). Here, the influence of the artificial light is considerably lower and the difference between the sensor values shrinks down to values around 0 at noon. Obviously, the artificial light in the room cannot add to the recorded sensor value anymore once a certain light level is reached in the environment. Therefore, the time period of the coordinate assignment must be selected carefully if nodes are placed at especially exposed locations. However, our experiences also show that for normal conditions event detection is possible all day if the event threshold is set to a reasonable value.

The last step of our evaluation is to investigate the results of using the third approach with different event detection thresholds. We performed a set of experiments choosing different event thresholds at different times of the day. In all of these experiments the client application sent the coordinate messages with the minimum possible transmission power and we used an event detection timer length of 4 seconds. Table 2 shows an overview of the results. Note that we deliberately ignored the RSSI values for this evaluation to emphasize the benefit of using events. However, the RSSI value could be used as an additional criterion.

If the event threshold is set to a too small value, then events can occur without being explicitly triggered by a user simply due to the variations of the sensor values over time. The consequence of this are so-called false positives

Table 2. Average success rates for different event thresholds

Event detection threshold	Avg. % of nodes assigned correctly
5	53.57
10	100.0
20	100.0
100	78.57

during the assignment of symbolic coordinates – nodes that assign a coordinate in reaction to a coordinate message without lying in the room where the event is actually triggered. In our experiments we could observe this for an event threshold of 5 (see Table 2). The results of experiments performed in the evening or at night with this threshold lay above the average but even then the artificial light oscillated enough to generate some false positives.

To explore the other end of the spectrum, i.e., a high event detection threshold, we performed experiments with a threshold value of 100. As expected, the high event detection threshold reliably prevented the occurrence of any false positives. However, some of the intended recipients also did not detect an event and consequently did not assign the symbolic coordinate resulting in a success rate well below 100%.

With an event threshold of 10 or 20 all of our experiments assigned the correct symbolic coordinates to all nodes in the network irrespective of the time of day the experiment was performed. Therefore, selecting an event threshold in this range provides for a reliable assignment of symbolic coordinates to sensor nodes.

Summarizing the results of our evaluation, assigning symbolic coordinates by broadcast has shown a good performance when used together with RSSI filtering. If, however, a high accuracy is required, then assigning symbolic coordinates by assisted broadcast is able to provide a reliable solution as has been shown by our experiments.

6   Conclusions and Future Work

In this paper we have motivated the benefit of using symbolic coordinates in wireless sensor networks and have discussed the advantages and disadvantages of two canonical approaches for the assignment of such coordinates – the individual assignment and the assignment by broadcast. We have then presented a third solution that combines the advantages of both approaches with only minor additional effort for an externally triggered sensor event. The evaluation shows that this approach allows a simple yet reliable assignment of symbolic coordinates to sensor nodes in indoor scenarios. This way, the manual configuration of symbolic coordinates after the deployment of a sensor network is a viable alternative to more sophisticated node localization approaches.

A reliable indoor localization system for sensor networks that is able to determine the position of nodes without requiring user interaction or expensive hardware while only generating a low message and computational overhead is definitely desirable. However, while this is not foreseeable, our solution provides a reliable assignment of symbolic coordinate information to sensor nodes that only generates a low overhead and only requires a reasonable amount of support by the user.

As part of future work we plan to investigate other types of sensor stimuli that can be used to disambiguate the assignment of symbolic coordinates to nodes. Particularly, we are interested in looking at combinations of sounders and microphones.

We are also interested in a combination of sensor networks and building automation systems, which automate the control of different mechanical and electrical systems in buildings. This would make it possible to turn on and turn off the lights in the rooms of a building automatically. This way, we could completely automate the assignment of symbolic coordinates by sending out symbolic coordinate messages and then triggering the light sensors in the respective rooms of the building.

Another aspect that we are actively working on is the grouping of nodes based on sensor data. Instead of relying on sensor stimuli deliberately triggered by the user we are interested in analyzing the sensor data collected by sensor nodes as part of their normal operation. Based on a similarity analysis of the sensor data from different nodes we want to decide which nodes reside together in the same room and group these nodes together.

References

1. Langendoen, K., Reijers, N.: Distributed localization in wireless sensor networks: a quantitative comparison. Computer Networks 43(4) (2003) 499–518
2. Priyantha, N.B., Chakraborty, A., Balakrishnan, H.: The cricket location-support system. In: Proceedings of the 6th Int. Conf. on Mobile computing and networking. (2000)
3. Elnahrawy, E., Li, X., Martin, R.P.: The limits of localization using signal strength: A comparative study. In: Proceedings of The First IEEE International Conference on Sensor and Ad hoc Communications and Networks (SECON 2004). (2004)
4. Stoleru, R., He, T., Stankovic, J.A., Luebke, D.: A high-accuracy, low-cost localization system for wireless sensor networks. In: SenSys '05: Proceedings of the 3rd international conference on Embedded networked sensor systems. (2005)
5. Corke, P., Peterson, R., Rus, D.: Networked robots: Flying robot navigation using a sensor net. In: Proceedings of the Eleventh International Symposium of Robotics Research (ISRR). (2003)
6. Brumitt, B., Shafer, S.: Topological world modeling using semantic spaces. In: Workshop Proc. of Ubicomp: Location Modeling for Ubiquitous Computing. (2001)
7. Jiang, C., Steenkiste, P.: A hybrid location model with a computable location identifier for ubiquitous computing. In: UbiComp '02: Proceedings of the 4th international conference on Ubiquitous Computing, London, UK, Springer-Verlag (2002) 246–263
8. Becker, C., Dürr, F.: On location models for ubiquitous computing. Personal Ubiquitous Computing 9(1) (2005) 20–31

Routing Strategies for Wireless Sensor Networks

Raúl Aquino-Santos[1], Luis Villasenor-Gonzalez[2], Jaime Sanchez[2], José Rosario Gallardo[2]

[1] Faculty of Telematics, University of Colima
28040, Av. Universidad 333, Colima, Colima, México
[2] CICESE, Research Centre
22860, Carretera Tijuana-Ensenada, k. 113
Emsenada, B.C.N., Mexico
aquinor@ucol.mx; {luisvi, jasan, jgallard@cicese.mx}

Abstract. This paper evaluates three routing strategies for wireless sensor networks: source, shortest path, and hierarchical-geographical, which are the three most commonly employed by wireless ad-hoc and sensor networks algorithms. Source routing was selected because it does not require costly topology maintenance, while shortest path routing was chosen because of its simple discovery routing approach and hierarchical-geographical routing was elected because it uses location information via Global Positioning System (GPS). The performance of these three routing strategies is evaluated by simulation using OPNET, in terms of latency, End to End Delay (EED), packet delivery ratio, routing overhead, overhead and routing load.

Keywords: Wireless sensor networks, multi-hop networks, unicast routing, hierarchical and flat routing mechanisms.

1 Introduction

Recent advances in micro-electro-mechanical systems (MEMS) technology have made the deployment of wireless sensor nodes a reality [1] [2], in part, because they are small, inexpensive and energy efficient. Each node of a sensor network consists of three basic subsystems: a sensor subsystem to monitor local environment parameters, a processing subsystem to give computation support to the node, and a communication subsystem to provide wireless communications to exchange information with neighboring nodes. Because each individual sensor node can only cover a relatively limited area, it needs to be connected with other nodes in a coordinated manner to form a sensor network (SN), which can provide large amounts of detailed information about a given geographic area. Consequently, a wireless sensor network (WSN) can be described as a collection of intercommunicated wireless sensor nodes which coordinate to perform a specific action. Unlike traditional wireless networks, WSNs depend on dense deployment and coordination to carry out their task. Wireless sensor nodes measure conditions in the environment surrounding them and then transform these measurements into signals that can be

Please use the following format when citing this chapter:

Aquino-Santos, R., Villasenor-Gonzalez, L., Sanchez, J., Gallardo, J. R., 2007, in IFIP International Federation for Information Processing, Volume 248, Wireless Sensor and Actor Networks, eds. L. Orozco-Barbosa, Olivares, T., Casado, R., Bermudez, A., (Boston: Springer), pp. 191-202.

processed to reveal specific information about phenomena located within a coverage area around these sensor nodes.

WSNs have a variety of applications. Examples include environmental monitoring – which involves monitoring air, soil and water, condition-based maintenance, habitat monitoring (determining the plant and animal species population and behavior), seismic detection, military surveillance, inventory tracking, smart spaces, etc. [3][4]. Despite their many diverse applications, WSNs pose a number of unique technical challenges due to the following factors: fault tolerance (robustness), scalability, production costs, operating environment, sensor network topology, hardware constraint, transmission media and power consumption.

To date, the ZigBee Alliance is developing a communication standard for WSNs to support low-cost, low-power consumption, two-way wireless communications. Solutions adopting the ZigBee standard will be embedded in consumer electronics, home and building automation, industrial controls, PC peripherals, medical sensor applications, toys and games [5].

Sensor networks are generally deployed into an unplanned infrastructure where there is no a priori knowledge of their specific location. The resulting problem of estimating the spatial coordinates of the node is referred to as location. Most of the proposed localization methods today depend on recursive trilateration/multilateration techniques [6].

In WSNs, obtaining data is sometimes more important than knowing the specific Id of the originating node. Because the data collected by many sensors in WSNs is typically based on a common phenomenon, there is a high probability that this data has some degree of redundancy. Data redundancy needs to be exploited by the routing protocol to optimize energy and bandwidth utilization.

Many researchers are currently engaged in developing strategies to meet these many diverse requirements. This paper focuses on a performance analysis of three basic routing strategies which are commonly used in routing protocols in wireless ad-hoc and sensor networks. The remainder of the paper is organized as follows: Section 2 considers various routing protocols that deal with state-of-the-art routing techniques for wireless sensor networks. Section 3 provides an explanation of the scenario simulated and finally, Section 4 summarizes our work and proposes future research.

2   State-of-the-art of Routing Techniques for Wireless Sensor Networks

Routing protocols for wireless sensor networks can be classified as data-centric, hierarchical or location-based [7]. In these three categories, source, shortest path, and hierarchical-geographical strategies play an important role to develop all of the routing protocols.

2.1 Data-centric protocols

In data-centric protocols, the sensor nodes broadcast an advertisement for the available data and wait for a request from an interested sink. Flooding is a simple technique that can be used to broadcast information in wireless sensor networks, however it requires significant resources because each node receiving a message must rebroadcast it, unless a maximum number of hops for the packet are reached, or the destination of the packet is the node itself. Flooding is a reactive technique that does not require costly topology maintenance or complex route discovery algorithms. However, it does have several additional deficiencies such as: implosion, overlap and resource blindness [8]. A derivation of flooding is gossiping, in which nodes do not broadcast. Instead, they send the incoming packets to a randomly selected neighbor.

Sensor protocols for information via negotiation (SPIN) address the deficiencies of classic flooding by providing negotiation and resource adaptation [9]. However, SPIN data advertisement mechanism cannot, by itself, guarantee data delivery [10]. SPIN employs a shortest path strategy based on three types of messages to communicate:

ADV – new data advertisement. When a SPIN node has data to share, it can advertise this fact by transmitting an ADV message containing meta-data.

REQ – request for data. A SPIN node sends an REQ message when it wishes to receive some actual data.

DATA – data message. DATA messages contain actual sensor data with a meta-data header.

Unlike traditional networks, a sensor node does not necessarily require an identity (e.g. an address). Instead, applications focus on the different data generated by the sensors. Because data is identified by its attributes, applications request data by matching certain attribute values. One of the most popular algorithms for data-centric protocols is direct diffusion and it bases its routing strategy on shortest path [11]. A sensor network based on direct diffusion exhibits the following properties: each sensor node names data that it generates with one or more attributes, other nodes may express interests based on these attributes, and network nodes propagate interests. Interests establish gradients that direct the diffusion of data. In its simple form, a gradient is a scalar quantity. Negative gradients inhibit the distribution of data along a particular path, and positive gradients encourage the transmission of data along the path.

The Energy-Aware Routing protocol is a destination-initiated reactive protocol that increases the network lifetime using only one path at all times, it seems very similar to source routing [12]. Rumor routing [13] is a variation of direct diffusion that is mainly intended for applications where geographic routing is not feasible. Gradient-based routing is another variant of direct diffusion [14]. The key idea of gradient-based routing is to memorize the number of hops when the interest is diffused throughout the network. Constraint Anisotropic Diffusion Routing (CADR) is a general form of direct diffusion [15] and lastly, Active Query Forwarding in Sensor Networks (ACQUIRE) [16] views the network as a distributed database, where complex queries can be further divided into several sub queries.

2.2 Hierarchical protocols

Hierarchical protocols are based on clusters because clusters can contribute to more scalable behavior as the number of nodes increases, provide improved robustness, and facilitate more efficient resource utilization for many distributed sensor coordination tasks.

Low-Energy Adaptive Clustering Hierarchy (LEACH) is a cluster-based protocol that minimizes energy dissipation in sensor networks by randomly selecting sensor nodes as cluster-heads [17]. Power-Efficient Gathering in Sensor Information System (PEGASIS) [18] is a near optimal chain-based protocol. The basic idea of the protocol is to extend network lifetime by allowing nodes to communicate exclusively with their closest neighbors, employing a turn-taking strategy to communicate with the Base Station (BS). Threshold-sensitive Energy Efficient protocol (TEEN) [19] and Adaptive Periodic TEEN (APTEEN) [20] have also been proposed for time-critical applications. In TEEN, sensor nodes continuously sense the medium, but data transmission is done less frequently. APTEEN, on the other hand, is a hybrid protocol that changes the periodicity or threshold values used in the TEEN protocol, according to user needs and the application type.

2.3 Location-based protocols

In location-based routing, the forwarding decision by a node is primarily based on the position of a packet's destination and the position of the node's immediate one-hop neighbor. The position of the destination is contained in the header of the packet. If a node has a more accurate position of the destination, it may choose to update the position in the packet before forwarding it. The position of the neighbors is typically learned through a one-hop broadcast beacon. These beacons are sent periodically by all nodes and contain the position of the sending node.

We can distinguish three main packet-forwarding strategies for position-based routing: greedy forwarding, restricted directional flooding, and hierarchical approaches. For the first two, a node forwards a given packet to one (greedy forwarding) or more (restricted directional flooding) one-hop neighbors that are located closer to the destination than the forwarding node itself. The selection of the neighbor in the greedy case depends on the optimization criteria of the algorithm. The third forwarding strategy is to form a hierarchy in order to scale to a large number of mobile nodes.

Minimum Energy Communication Network (MECN) [21] establishes and maintains a minimum energy network for wireless networks by utilizing low-power geographic positioning system (GPS). The main idea of MECN is to find the sub-network with the smallest number of nodes that requires the least transmission power between any two particular nodes (shortest path). The Small Minimum Energy Communication Network (SMECN) [22] is an extension of MECN. The major drawback with MECN is that it assumes every node can transmit to every other node, which is not always possible. One advantage of SMECN is that it considers obstacles between pairs of nodes. Geographic Adaptive Fidelity (GAF) [23] is an energy-aware location-based routing algorithm primarily designed for ad-hoc networks that can also be applied to sensor networks. GAF conserves energy by turning off unnecessary nodes in the network without affecting the level of routing fidelity. Finally, Geographic and Energy Aware Routing [24] uses energy-awareness and geographically informed neighbor selection heuristics to route a packet toward the destination region.

2.4 ZigBee Protocol

The IEEE 802.15.4-2003 standard defines the lower two layers: the physical (PHY) layer and the medium access control (MAC) sub-layer. The ZigBee alliance builds on this foundation by providing the network (NWK) layer and the framework for the application layer, which includes the application support sub-layer (APS), the ZigBee device objects (ZDO) and the manufacturer-defined application objects.

IEEE 802.15.4-2003 has two PHY layers that operate in two separate frequency ranges: 868/915 MHz and 2.4 GHz. The 2.4 GHz mode specifies a Spread Spectrum modulation technique with processing gain equal to 32. It handles a data rate of 250 kbps, with Offset-QPSK modulation, and a chip rate of 2 Mcps.

The 868/915 MHz mode specifies a DSSS modulation technique with data rates of 20/40 kbps and chip rates of 300/600 kcps. The digital modulation is BPSK and the processing gain is equal to 15.

On the other hand, the MAC sub-layer controls access to the radio channel using a CSMA-CA mechanism. Its responsibilities may also include transmitting beacon frames, synchronizing transmissions and providing a reliable transmission mechanism.

The responsibilities of the ZigBee NWK layer includes mechanisms used to join and exit a network, in order to apply security to frames and to route frames to their intended destinations based on shortest path strategy. In addition, the discovery and maintenance of routes between devices transfer to the NWK layer. Also, the discovery of one-hop neighbors and the storing of pertinent neighbor information are done at the NWK layer. The NWK layer of a ZigBee coordinator is responsible for starting a new network, when appropriate, and assigning addresses to newly associated devices.

The responsibilities of the APS sub-layer include maintaining tables for binding, which is the ability to match two devices together based by their services and their needs, and forwarding messages between bound devices. The responsibilities of the ZDO include defining the role of the device within the network, initiating and/or responding to binding requests and establishing a secure relationship between network devices. The ZDO is also responsible for discovering devices on the network and determining which application services they provide.

3 Scenarios Simulated

The routing protocols described in section 2 make use of one, or a combination of the following strategies: source, shortest path or hierarchical-geographical routing strategies. The performance of these basic strategies is evaluated using the following metrics:

- Route discovery time (Latency): is the time the sink has to wait before actually receiving the first data packet.
- Average end-to-end delay of data packets: are all possible delays caused by queuing, retransmission delays at the MAC and propagation and transfer times.
- Packet delivery ratio: is the ratio of the number of data packets delivered to the destination and the number of data packets sent by the transmitter. Data packets may be dropped en route for several reasons: e.g. the next hop link is broken when the data packet is ready to be transmitted or one or more collisions have occurred.
- Routing load: is measured in terms of routing packets transmitted per data packets transmitted. The latter includes only the data packets finally delivered at the destination and not the ones that are dropped. The transmission at each hop is counted once for both routing and data packets. This provides an idea of network bandwidth consumed by routing packets with respect to "useful" data packets.
- Routing overhead: is the total number of routing packets transmitted during the simulation. For packets sent over multiple hops, each packet transmission (hop) counts as one transmission.
- Overhead (packets): is the total number of routing packets generated divided by the sum of total number of data packets transmitted and the total number of routing packets

3.1 Basic Routing Strategies implemented

In source routing, each packet header carries the complete ordered list of nodes through which the packet must pass. The key advantage of source routing is that intermediate nodes do not need to maintain up-to-date routing information in order to route the packets they forward, since the packets themselves already contain all the routing information. This fact, coupled with the on-demand nature of the protocol, eliminates the need for the periodic route advertisement and neighbor detection packets present in other protocols such as the Energy Aware Routing.

In the shortest path strategy, when a node S needs a route to destination D, it broadcasts a route request message to its neighbors, including the last known sequence number for that destination. The route request is flooded in a controlled manner through the network until it reaches a node that has a route to the destination.

Each node that forwards the route request creates a reverse route for itself back to node S. Examples are SPIN, Direct Diffusion, MECN, and the ZigBee standard.

When the route request reaches a node with a route to D, that node generates a route reply containing the number of hops necessary to reach D and the sequence number for D most recently seen by the node generating the reply. Importantly, each node that forwards this reply back toward the originator of the route request (node S) creates a forward route to D. The state created in each node remembers only the next hop and not the entire route, as would be done in source routing.

Hierarchical-geographical strategy improves the traditional routing strategies based on non-positional routing by making use of location information provided by GPS as it minimizes flooding of its Location Request (LREQ) packets. Flooding, therefore, is directive for traffic control by using only the selected nodes, called gateway nodes to diffuse LREQ messages. The purpose of gateway nodes is to minimize the flooding of broadcast messages in the network by reducing duplicate retransmissions in the same region.

Member nodes are converted into gateways when they receive messages from more than one cluster-head. All the members of the cluster read and process the packet, but do not retransmit the broadcast message. This technique significantly reduces the number of retransmissions in a flooding or broadcast procedure in dense networks. Therefore, only the gateway nodes retransmit packets between clusters (hierarchical organization). Moreover, gateways only retransmit a packet from one gateway to another in order to minimize unnecessary retransmissions, and only if the gateway belongs to a different cluster-head.

We decided to evaluate source, shortest path and hierarchical-geographical routing strategies since they represent the foundation of all of the above mentioned routing protocols.

The simulator for evaluating the three routing strategies for our wireless sensor network is implemented in OPNET 11.5, and the simulation models a network of 225 MICAz sensor nodes [2]. This configuration represents a typical scenario where nodes are uniformly placed within an area of 1.5 km2.

We used a 2405- 2480 MHz frequency range and a 250 kbps data rate for our simulation, with a MICAz sensor node separation of 75 m. This scenario represents a typical wireless sensor network with one sink node acting as a gateway to communicate the WSN with a separate network (Internet). In our scenario one sensor node communicates with the sink, and the sensor node sends a packet every second (constant bit rate).

3.2 Simulation Results

Figure 1 shows the latency between the sink and the source in milliseconds. Source and shortest path routing strategies show a similar behavior. However, hierarchical-geographical routing shows the poorest behavior due to the transmission of position information via hello packets which produce more collision in the wireless medium, in addition, the cluster formation mechanism also increase the latency.

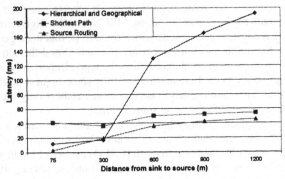

Fig. 1. Latency (milliseconds).

Figure 2 shows the End-to-End Delay (EED) between the sink and the source in milliseconds. The hierarchical-geographical routing strategy shows the worst behavior because the static nature of the wireless sensor nodes causes synchronization of the packets. Synchronization arises from the simultaneous transmission of packets between neighbors. As results, the frequent transmission of Hello packets produces more collision with data packets.

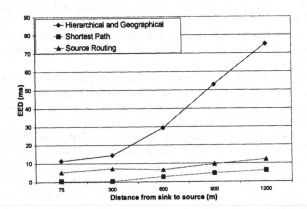

Fig. 2. End-to-End Delay (milliseconds).

The three routing mechanisms show a similar behavior in terms of packet delivery ratio because of their static nature, as illustrated in figure 3.

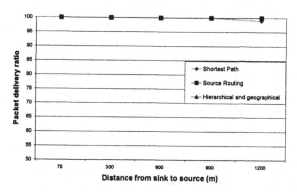

Fig. 3. Packet delivery ratio.

Figure 4 shows the Routing Overhead between the sink and the source. Routing overhead is the total number of routing packets transmitted during the simulation. Again, the shortest path routing strategy performs the best and the hierarchical-geographical strategy the worst. This is due to the Hello packets used for the cluster formation mechanism.

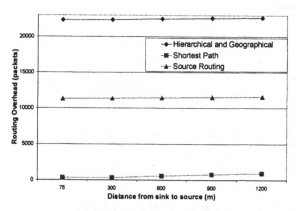

Fig. 4. Routing Overhead (packets).

Figure 5 shows the overhead between the sink and the source. The shortest path technique also has the best performance, with source routing and hierarchical-geographical mechanism performing in a similar fashion.

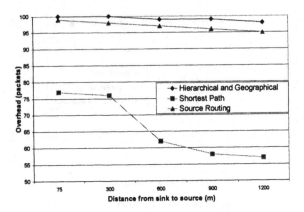

Fig. 5. Overhead (packets).

Figure 6 shows the Routing Load between the sink and the source. This metric provides an idea of how much network bandwidth is consumed by routing packets in relation to the useful data packets actually received. Once again, the shortest path routing strategy has the best performance, and the hierarchical-geographical mechanism the worst.

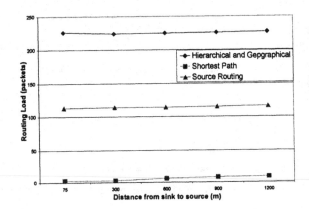

Fig. 6. Routing Load (packets).

4   Conclusions and Future Work

In this paper, we have evaluated three basic routing strategies widely used in routing protocols for wireless sensor networks. Source routing only improves shortest path and hierarchical-geographical routing in terms of latency. The main disadvantage of source routing is that it lacks a number of hops metric, which can frequently result in longer path selection. Shortest path behaves well in terms of EED, routing overhead, overhead and routing load. Hierarchical-geographical routing performs the worst because it must send hello packets in order to acquire and transmit location information. This consideration makes hierarchical-geographical routing in wireless sensor networks more weighty because it transmits hello packets more frequently, requiring greater bandwidth and energy resources. However, despite these significant disadvantages, hierarchical-geographical routing remains the routing option most often used in health, military, agriculture, robotic, environmental and structural monitoring. An important area of future research is to optimize hierarchical-geographical routing algorithm to facilitate its use in large geographical areas requiring dense sensor distribution.

Acknowledgements

The authors acknowledge the support received by the National Council of Science and Technology (CONACYT) under project grant No. 48391.

References

1. V. Rajaravivarma, Yi Yang, and Teng Yang. An Overview of Wireless Senor Network and Applications. Proceeding of the 35th Southeastern Symposium on System Theory, pp. 432-436, 2003.
2. http://www.xbow.com/Products/Wireless_Sensor_Networks.htm
3. Stephan Olariu and Qingwen Xu. Information Assurance in Wireless Sensor Networks. Proceedings of the 19th IEEE International Parallel and Distributed Processing Symposium, pp. 236 -240, 2005.
4. Alan Mainwaring, Joseph Polastre, Robert Szewczyk, David Culler, John Anderson. Wireless Sensor Networks for Habitat Monitoring. Proceeding of the 1st ACM International workshop on wireless sensor Networks and applications, pp. 88-97, 2002.
5. ZIgBee Specification. ZigBee Document 053474r06, version 1.0. December 2004.
6. Nirupama Bulusu, John Heidemann, Deborah Strin. GPS-less Low Cost Outdoor Localization For Very Small Devices. IEEE Personal Communication, vol. 7, issue 5, pp. 28-34, 2000.
7. A Reliable Routing Protocol Design for wireless sensor Networks. Yanjun Li, Jiming Chen, Ruizhong Lin, Zhi Wang. International Conference on Mobile Ad-hoc and Sensor Systems, 2005.
8. I.F. Akyildiz, W. Su, Y. Sankarasubramaniam, E. Cayirci. Wireless sensor networks: a survey. Computer Networks, pp. 393-422, 2002.
9. Wendi Rabiner Heinzelman, Joana Kulik, and Hari Balakrishnan. Adaptive Protocols for Information Dissemination in Wireless Sensor Networks. Proceedings of the 5th annual ACM/IEEE International Conference on Mobil Computing and Networking (MOBICOM), pp. 174-185, 1999.
10. Jamal N. Al-karaki, Ahmed E. Kamal. Routing Techniques in Wireless Sensor Networks: A survey. IEEE Wireless Communications, pp. 6-28, 2004.

11.Deborah Estrin, Ramesh Govindan, John Heidemann, Satish Kumar. Next Century Challenges: Scalable Coordination in Sensor Networks. Proceedings of the 5th ACM/IEEE International Conference on Mobile Computing and Networking, pp. 263-270, 1999.

12.Shah, R. C. Rabaey, J. M. Energy Aware Routing for low Ad Hoc Sensor Networks. IEEE Wireless Comunications and Networks Conference, vol. 1, pp. 350-355, 2002.

13.David Braginsky, Deborah Estrin. Rumor Routing Algorithm for Sensor Netorks. International Conference on Distributed Computing Systems (ICDCS-22), 2002.

14.Curt Schurgers, Mani B. Srivastava. Energy Efficient Routing in Wireless Sensor Networks. Proceeding of the Communication for Network-centric operations: creating the information force, 2001.

15.Maurice Chu, Horst Haussecker, and Feng Zhao. Scalable Information-Driven Sensor Querying and Routing for ad hoc Heterogeneous Sensor Networks. International Journal of High Performance Computing Applications, 2002.

16.Narayanan Sadagopan, Bhaskar Krishnamachari, and Ahmed Helmy. The ACQUIRE Mechanism for Efficient Querying in Sensor Networks. Proceedings of the IEEE International Workshop on Sensor Network Protocols and Applications (SNPA), in conjunction with IEEE ICC, pp. 149-155, 2003.

17.Heinzelman, W. R. Chandrakasan, A. Balakrishnan, H. Energy-efficient communication protocol for wireless microsensor networks. Proceedings of the 33rd Annual Hawaii International Conference on System Sciences, vol. 2, pp. 1-10, 2000.

18.Stephanie Lindsey and Cauligi S. Raghavendra. PEGASIS: Power-Efficient GAthering in Sensor Information Systems. Proceeding of the IEEE Aerospace Conference, vol. 3, pp. 1125-1130, 2002.

19.Arati Manjeshwar and Dharma P. Agrawal. TEEN: A Routing Protocol for Enhanced Efficiency in Wireless Sensor Networks. Proceedings of the 15th International Symposium on Parallel and Distributed Processing, pp. 2009-2015, 2001.

20.Arati Manjeshwar and Dharma P. Agrawal. APTEEN: A Hybrid Protocol for Efficient Routing and Comprehensive Information Retrieval in Wireless Sensor Networks, Proceedings of the 16th International Symposium on Parallel and Distributed Processing, pp.195-202, 2002.

21.Volkan Rodoplu and Teresa H. Meng. Minimum Energy Mobile Wireless Networks. IEEE Journal on selected areas in communications, vol. 17, issue 8, pp. 1333-1344, 1999.

22.Li Li, Joseph Y. Halpern. Minimum-Energy Mobile Wireless Networks Revisited. IEEE International Conference on Communications, vol. 1, pp. 278-283, 2001.

23.Ya Xu, John Heideman, and Deborah Estrin. Geography-informed Energy Conservation for Ad-Hoc Routing. In proceedings of the ACM/IEEE International Conference on Mobile Computing and Networking, pp. 70-84, 2001.

24.Yan Yu, Ramesh Govindan, Deborah Estrin. Geographic and Energy Aware Routing: a recursive data dissemination protocol for wireless sensor networks. UCLA Computer Science Department Technical Report UCLA/CSD-TR-01-0023, 2001.

Hierarchical Geographic Routing for Wireless Ad-Hoc Networks

Luis A. Hernando, Unai Arronategui

I3A, University of Zaragoza
C/ María de Luna 1, Ed. Ada Byron, 50018 Zaragoza, Spain
lahernan@unizar.es, unai@unizar.es

Abstract. Geographic routing is a well established solution for scaling in large wireless ad-hoc networks. A fundamental issue is forwarding packets around voids to reach destinations in networks with sparse topologies. All general known solutions need first to get into a dead end, at link level, to be able afterwards to apply a recovery algorithm. These techniques can lead to very inefficient forwarding paths. We propose a novel general approach, based on light weight connectivity maps, C-Maps, distributed among all the nodes in the network to obtain more efficient and robust paths. The main contribution of our method is the distributed Mercator protocol that builds these maps. Each node in this protocol builds and maintains its own C-Map that summarizes connectivity information of all the network around itself using hierarchical regions. This information is more precise from regions closer to the node. Nodes apply greedy forwarding and face routing to the different hierarchical levels of connectivity information. Better paths are obtained with this behavior. Robustness is guaranteed by every node containing its C-Map. Our analytical and simulation work shows that the map state and the communication overhead grows logarithmically with the size of the network.

Key words: Hierarchical Geographic Routing, Wireless Ad Hoc Networks, Connectivity Maps.

1 Introduction

Network scalability is a fundamental issue in routing protocols for wireless ad-hoc networks. An attractive approach to this problem is geographic routing [2, 7] because no other routing information than the set of neighbours has to be maintained. Also, in dense networks, the forwarding method used, greedy forwarding, is very simple. Besides, cross layer methods allow efficient data-centric applications [11] using geographic routing algorithms.

If the network is not dense enough, packets can find voids that might block greedy forwarding with dead ends and, thus, packets can not reach destinations in a network where connectivity exists. The traditional solution to get out of the dead end and forward around the void is face routing [6], which applies the

Please use the following format when citing this chapter:

Hernando, L. A., Arronategui, U., 2007, in IFIP International Federation for Information Processing, Volume 248, Wireless Sensor and Actor Networks, eds. L. Orozco-Barbosa, Olivares, T., Casado, R., Bermudez, A., (Boston: Springer), pp. 203-214.

"right hand rule" in a planar subgraph of the network. But, all general geographic routing algorithms needs to get first in these dead ends at the link level so that, afterwards, they can apply a recovery algorithm to forward around the void [9]. And, in this way, very inefficient paths can be obtained, mainly when dealing with big voids and, worse, with big and/or frequent concavities at the borders of those voids.

We propose a novel general approach to avoid arriving at dead ends at link level and getting more efficient paths to the destination node. Each node builds and maintains a contextual map, a *C-Map*, containing synthesized information of connectivity of hierarchical regions surrounding the node. These regions are square tiles obtained from simple geographical operations.

We assume that radio ranges could be non uniform, links are bi-directional and nodes are assigned absolute coordinates in the network.

Network connectivity information is folded in region connectivity information at the different levels of the map hierarchy. Finally greedy forwarding and face routing is applied on the resulting region connectivity graph. Using geographic routing at region level we achieve more efficient paths, mainly in sparse configuration with big voids and concavities.

Also, node mobility can easily be included integrating the proposal in [10] with our techniques. Energy considerations are not covered in this paper.

The rest of the paper is organized as follows. Section 2 shows a review of existing and related work. In section 3, the connectivity maps are described. Section 4 presents Mercator, the maps construction protocol. Section 5 presents the hierarchical routing protocol. Section 6 offers theoretical information about costs induced by this method. In section 7, simulation results are shown. And, finally, section 8 offers the conclusions of this work.

2   Related Work

GPSR [4] was the first geographic routing algorithm that tackled the problem of voids and concavities. Messages are forwarded using greedy mode whenever possible towards destination until they encounter a dead end. Then GPSR switches to perimeter mode, the recovery algorithm, forwarding the message with a right-hand rule algorithm over a planarized graph outside the problematic area, then, GPSR switches back to greedy mode. GOAFR family of algorithms [7] also implements face routing over a planarized graph as recovery algorithm, improving GPSR's efficiency. It was proved in [5] that geographic routing protocols (like GPSR or GOAFR) which took the *unit disk graph* assumption do not behave correctly. CLDP [5] provides a planarized graph without relying on the previous assumption by continuously probing all available links, which results in a high cost.

GDSTR [9] was proposed as an alternative: instead of using expensive graph planarization algorithms, it maintains a spanning tree, the *Hull Tree*. Nodes in the Hull Tree keep notion of the area covered by nodes in the branches down to the leafs. GDSTR's recovery algorithm consists in forwarding messages towards the tree's root until they are outside the problematic area.

All these methods need to get into a dead end to be able to apply a recovery algorithm. Depending on the shape of each void, mainly on its size and concavities existence, very long and inefficient paths may result.

In [3] it is argued that to avoid getting into dead ends is better solution that trying to apply recovery algorithms afterwards. The Distance Upgrading algorithm is developed to obtain better paths. But, this method is proposed only in the context of communications from sensors to a base station. Thus, only one destination is present and all other nodes calculate the distance to this unique destination.

Terminode routing [1] is a hierarchical routing protocol that uses geographic routing for distant destinations and more traditional link state routing in near neighbourhood. Terminode routing has a method called Geographical Map-Based Path Discovery for remote routing. It only operates to get anchored paths at the same level and it is assumed that a density map is already available somewhere outside the network. No distribution technique is presented for the maps. This trend is developed with the use of geographic maps [8] from vehicular navigation systems that are introduced in the geographic routing.

No method has been proposed where the network itself builds an explicit geographic connectivity map to improve substantially routing paths as we do.

3   Connectivity Maps

Connectivity information between different areas in the network is stored into a map, the C-Map. The network is modeled as a hierarchy of square areas defined on a 2D euclidean space. Nodes obtain its localization coordinates by means of an external positioning system like GPS or other localization techniques.

A C-Map stores information about a hierarchy of nested square regions surrounding the owner node's location. The highest level of the map, with the biggest squares, determines the area covered by the whole map. Lower levels keep information about connectivity of small areas near the node, while higher levels keep summarized connectivity information about large and faraway areas.

Hierarchy of network regions. Information in a C-Map is organized into M *Map Levels*, starting from 0, the lowest one, to $M - 1$. Each map level is composed of Q non-overlapping squares known as *tiles*, properly sized to fit four tiles of level m into one of level $m + 1$. The map is centered in order to balance the amount of information in all directions around the node and to facilitate the composition of information explained later, as shown in Fig. 1. Tiles have to be carefully placed according to the following rules:

- Map level m must be completely nested into map level $m + 1$, and thus $Q = 4^n, n \in \mathbb{N}$.
- In each Map level, the tile containing the owner node is known as *central tile* and must be completely surrounded by other tiles.

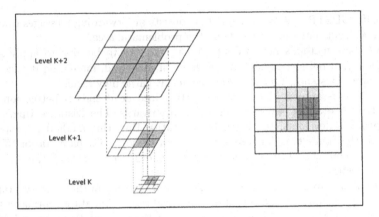

Fig. 1. Hierarchy of nested levels centered around the node owner of the C-Map

Connectivity between regions. Connectivity is defined as the possibility for a message coming from a tile to reach a contiguous one across their common border. Each tile stores the connectivity information of its corresponding square region of the network with its surrounding squares.

Having the connectivity information about all the tiles of each map level, the owner of the map is able to quickly find out if a region of the network is reachable across a path of connected tiles.

Connectivity inside tiles. At first, a tile marked as reachable through all of its borders, would seem like an ideal tile, allowing communications traverse the tile completely, but it is not always true: nodes inside a tile might be divided into different unconnected groups known as *fragments*. This important issue implies that a fragmented tile is not always traversable from one border to other.

A fragment is defined as a connected group of nodes inside a tile and is always traversable.

4   The Mercator Protocol: C-Map Construction

The **Mercator** protocol builds, distributes and maintains the C-Maps for every node in the network in a fully decentralized way. All the information the protocol produces is based on basic connectivity status between nodes. First, connectivity between the smallest tiles is discovered with an interchange of HELLO messages. Then, that information is distributed to the immediate neighbour areas, giving nodes some knowledge about its vicinity and the ability to summarize the information they know. In subsequent iterations of this process, basic and summarized connectivity information will be shared using MAP messages, extending the ability of nodes to build information at higher levels.

After some time, all nodes will be provided with a map of its neighbourhood composed of levels at different scales. Later updates of the map will be required to address the topology changes produced by joining or exiting nodes or even entire network areas.

4.1   Previous Considerations

A square of level m, S_m, is identified by a tuple $<m, (i, j)>$ where (i, j) are S_m's grid coordinates using columns and rows. S_m's side length, L_m, is calculated as $L_m = L_0 * 2^m$, being L_0 a parameter of the network. Given a point P placed inside S_m, (i, j) can be calculated as the integer division of P's coordinates by L_m.

4.2   Mercator Protocol's Information: The C-Map

Information produced by **Mercator** protocol is stored into a C-Map at each node. A C-Map is composed of M Map levels. M is chosen according to the maximum diameter of the network. Each map level is composed of exactly $Q = 16$ tiles arranged in a 4x4 matrix layout (See Fig. 1). The map is 'centered' at the owner node's location, following the rules described in Sec. 3. When a node moves from one tile to another, its map has to be re-centered repositioning the tiles at all levels as needed to meet the previous rules.

As explained before, nodes inside a tile can be divided into fragments. Information stored in the C-Map about each tile comprehends the different fragments the tile is divided into, including for each one a representative identifier unique inside the tile and its connectivity status with neighbour fragments.

4.3   Discovery of Level 0 Information

Once the C-Map is centered, HELLO messages are broadcasted periodically by **Mercator** protocol to gather information about the fragment the node belongs to and its surroundings at the lowest level (level 0). HELLO messages are used also to maintain a fresh list of neighbour nodes.

Each HELLO message contains the following information: sender's node identifier and coordinates, sender's fragment identifier, fragments at neighbour tiles accessible from sender's fragment and the minimum number of hops needed to reach those neighbour fragments from the sender node.

Upon a reception of a HELLO message each node will update its state adding the new information: if HELLO messages are received from a node in a neighbour tile, then the shared border's status among both tiles is considered *active*. The sender's node fragment will be considered accessible from the receiver's fragment and hop count towards sender's fragment will be set to 0. By contrast, if a HELLO message is received from a node which is located in the same tile, then sender and receiver belong to the same fragment. Receiver node will check the HELLO message looking for information updates, i.e. a new active border or a new link with a fragment, and will keep this information that will be broadcasted in the next HELLO message.

After some stabilization time, each node will be able to compute its fragment's identifier based on which tile's borders seem active.

4.4   C-Map Addition Operation

Connectivity information is shared between nodes exchanging their C-Maps. The previous construction rules of C-Maps imply neighbours' C-Maps must be partially or totally overlapped. A node acquires information from its neigbours' C-Maps applying the addition operation.

Binary addition operator is defined for C-Maps: let C_1, C_2, C_r be C-Maps,

$$C_1 + C_2 \rightarrow C_r$$

where C_r is the combination of C_1 and C_2. C-Map addition operator works as follows: first, C_r is centered at the same point as C_1, then addition operator will select tiles from C_1 not present in C_2 and will copy its information (fragments) into C_r. Then it will select those tiles present in both maps and will add to C_r all fragments from C_1 as well as those from C_2. Finally, if central tiles of corresponding levels in C_1 and C_2 are neighbours and thus share a border, C_2 owner's fragment is marked as connected with C_1 owner's fragment into C_r.

4.5   C-Map Information Exchange

After some stabilization time, each node is able to provide information about its fragment of level 0 and its connections to other neighbour fragments. The node will then start broadcasting MAP messages to its immediate neighbours in order to share information about local and surrounding tiles. Initially, all 16 tiles of each level will be empty except the central tile of level 0, which will include the owner's node fragment.

Each MAP message carries the C-Map stored by the node at the sending time and its owner node's coordinates.

Upon a reception of a MAP message, receiver node will add the received C-Map to its own, acquiring the new information (as explained in 4.4). In the subsequent sendings, the new information will also be spread producing a propagation effect.

4.6   Higher Level Fragment Information Composition

Each node is in charge of the fragments it belongs to at each level and will build and keep them stored into its map. Each node computes from information about fragments of level $m - 1$ which fragment of level m it belongs to. Then MAP messages exchange will propagate fragment's information over the surrounding areas.

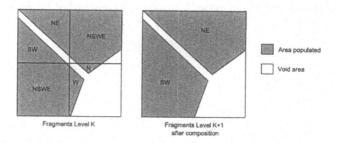

Fragments Level K   Fragments Level K+1
                                    after composition

Fig. 2. Composition of a higher level fragment. In the picture, fragments are identified with the cardinal directions of the tile's active borders for the fragment.

Levels in a C-Map are centered at the owner node's location (see 4.2), ensuring that all 16 tiles of level $m - 1$ fit perfectly into 4 tiles of level m. The composition algorithm computes node's fragment into its tile of level m, by examining the fragments in tiles of level $m - 1$ which fit into its tile of level m (see Fig. 2).

The composition algorithm works as follows: first, it looks for all fragments in the 4 tiles of level $m - 1$ which are connected (directly or by intermediate fragments) to the node's fragment at that level. Then, all those connected fragments will conform the node's fragment at level m, calculating also its identifier. Finally, the composed fragment is stored into the level m of the C-Map.

4.7 Information Expiring Mechanisms

Since ad-hoc networks are not static, nodes appear and disappear creating and removing routes, modifying the network's topology and connecting and disconnecting entire regions. For this reason an invalidation mechanism for obsolete information is proposed.

Every item of a C-Map, like a fragment inside a tile, is associated to an expiration timer that monotonically increases up to a maximum value. Timers information is also stored in the map. An item is removed from the C-Map when its associated timer reaches a predefined maximum value. Nodes in charge of an item produce the information about it, and periodically reset its timer value to 0 like a heartbeat. If an item of a map disappears from the network, i.e. a fragment gains some connectivity and it is renamed, nodes in charge of it will stop resetting the timer for the old identifier and after a period of time it will be removed from all C-Maps where it was contained.

4.8 A Global View of Built C-Maps

The behavior of the **Mercator** protocol on a network of about 500 nodes is shown in Fig. 3(a). It can be seen in Figs. 3(b),3(c) and 3(d) how connectivity information reflects the shape of the network including void areas and concavities and how a summarized network topology is constructed for each level.

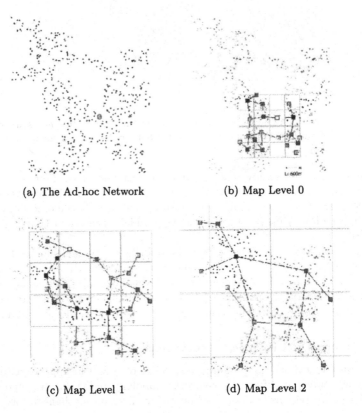

(a) The Ad-hoc Network   (b) Map Level 0

(c) Map Level 1   (d) Map Level 2

Fig. 3. C-Map Levels 0, 1 and 2 of the highlighted node in 3(a). Nodes of the same color inside a tile belong to the same fragment, which is represented by a colored square. Connectivity between fragments is represented with straight lines.

5   Hierarchical Geographic Routing with C-Maps

Having the C-Map, traditional geographic routing algorithms can be enhanced to take advantage of the available connectivity information at different scales. Big concavities and voids are very difficult to discover at link level, the lowest level in the map, but its easier to find them on the appropriate higher level.

An improved geographic routing algorithm that takes advantage of C-Maps would be pretty similar to the current ones, performing greedy forwarding through safe locations in the map, and thus, avoiding problematic areas in order of importance, instead of simply forwarding towards the destination point as in traditional geographic routing.

Enhanced Geographic Routing with C-Maps would work as follows: when a message is forwarded, a node would look in the map for the highest level where destination and current locations are placed inside different tiles of the Map Level. A greedy algorithm applied to the connectivity graph represented at that level will select the next fragment the message should be forwarded to. A waypoint is established at that fragment and the routing protocol will apply greedy forwarding towards that waypoint in the immediate lower level of the map. This process is repeated until Level 0, resulting in a direction suitable for greedy routing at link level. Known recovery algorithms would work in a similar way but applied to the graph at the level which encounters trouble.

With this approach problematic areas are avoided at all levels reducing to the minimum the usage of recovery algorithms.

6   Costs Estimation

The **Mercator** protocol has been designed in order to achieve high scalability properties with target networks from a few to hundred thousand nodes. The design has been focused on keeping conservative storage requirements per node and shared channel bandwidth usage.

C-Map storage cost. A look to the data structure managed by **Mercator** Protocol, the C-Map, shows the powerful scalability properties it offers. Each node in the network stores only its own C-Map which is composed of M Map Levels. Each Map Level stores $Q = 16$ tiles which might be divided into different fragments. Storage cost of a C-Map, S_{Cmap} can be calculated as follows:

$$S_{Cmap} = M * Q * F * f \tag{1}$$

Where F is the average number of fragments per tile and f is the storage cost of a fragment. S_{Cmap} increases linearly with M.

The central tile of each level is surrounded by at least one tile on each direction (See Sec. 3). This property guarantees for a Map Level K, a complete coverage of a network of L_K meters of diameter in a worst case scenario, where the owner node is in the farthest possible location from the middle point of the Map Level. The maximum coverage area of a C-Map, A_{Cmap}, is determined by its highest Map Level, in fact, M is chosen to guarantee coverage for a maximum network diameter. A_{Cmap} can be calculated as follows:

$$A_{Cmap} = (L_0 * \sqrt{Q} * 2^{M-1})^2 \tag{2}$$

Table 1 shows the coverage areas of C-Maps with different number of levels in a default setup, being $Q = 16$ and $L_0 = 500$ meters. A_{Cmap} increases exponentially with M.

M:	1	2	3	4	5	6	7	8
$A_{Cmap}(Km^2)$:	4	16	64	256	1024	4096	16384	65536

Table 1. Coverage area of a C-Map with M map levels

Mercator bandwidth usage. All **Mercator** information exchange is performed by means of local broadcast operations. Once the sender node transmits a message, it is not forwarded again by any of the receiver nodes.

Connectivity information discovery and distribution is done by means of periodic HELLO and MAP messages sendings. HELLO and MAP messages sending rate is calculated in order to use a low percent of the available channel bandwidth. Shared medium bandwidth used by *Mercator* in nodes surrounding point P, $M_{Bw}(P)$, can be approximated as follows:

$$M_{Bw}(P) = D(P) * A_{RR}(P) * (R_{HELLO} * S_{HELLO} + R_{MAP} * S_{MAP}) \quad (3)$$

Where, $D(P)$ is the average network population at point P, $A_{RR}(P)$ is the radio range area of a node in P, R_{HELLO} and R_{MAP} are the broadcast rates for HELLO and MAP messages, S_{HELLO} and S_{MAP} are the average size of HELLO and MAP messages and C_{Bw} is the channel bandwidth.

Wireless communications use a shared medium, thus, node density will affect the bandwidth employed by **Mercator** at any point in the network. We propose two strategies for dealing with node density:

1. Sending rates depend on D. For instance, taking $R_{MAP}(D) = K/D$, where K is a predefined constant, then, $R_{MAP}(D)$ will decrease as node density grows. This approach would help to reduce the channel bandwidth used by **Mercator**, turning M_{Bw} not dependent on D. Each node approximates D as the number of neighbours +1 divided by its area of radio coverage. This approach will produce an approximately constant bandwidth usage per square meter not dependent on D.
2. Use a notification mechanism instead of expiration timers. This mechanism, upon an inconsistency between information at a level with its immediate upper level would recalculate connectivity information for the upper level and spread it.

7   Experimental Results

We have implemented an ad-hoc network simulator to test the **Mercator** protocol under different scenarios. The following settings have been used in our tests: radio range is 200 meters. All nodes receive broadcasted messages within their radio range. R_{HELLO} is set to 1 message per second. R_{MAP} is set to 1/3 messages per second. Expiration ages are set to 4 *ticks* for each item. Tick duration is $1/R_{MAP} * 2^n$ seconds where n is the level number of the item monitored by the timer.

The side's length of a tile at level 0, L_0, is closely related to expiration ages and radio range. L_0 value has been set to 500 meters, which allows information to traverse an entire tile of level 0 in less than 4 hops without expiring.

Storage and bandwidth requirements: with the above parameters, a C-Map with 9 levels (262,000 square kilometers of coverage area and average network population of 2,600,000 nodes) requires 5.3KBytes of storage (according to (1)). In this scenario, an average neighbour count of 4 produces an approximated channel bandwidth usage of 71kbps (according to (3)).

Stabilization time. In a first test we have measured the map's stabilization time, which provides an idea of the response speed that the protocol offers against network topology changes. Stabilization time is measured by the number of MAP messages each node sends until the entire network map is completely built. We have simulated a cold startup in a network with 7000 nodes under different population density conditions (from 4 to 9 average neighbour count). As shown in Fig. 4, stabilization time increases linearly with the extension of the mapped area, and thus, exponentially with the number of levels of the C-Map. High density conditions help to increase information propagation speed and to reduce stabilization time.

We have also simulated dynamic scenarios with nodes joining and going away from the network. If a new node does not produce a substantial change into the network topology, stabilization time is about 1 message, by contrast, if a new link is created between two fragments, the stabilization time is similar to the case of a cold startup at the level of those fragments.

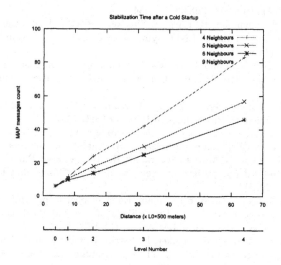

Fig. 4. Stabilization time after a cold startup

8   Conclusions

In this paper we propose a new approach in geographic routing that improves considerably routing paths in sparse networks with big voids and/or significant concavities. More efficient paths than classical routing and recovery algorithm can be obtained. Connectivity robustness is guaranteed with map distribution among all nodes. It can marry nicely with existing mobility solutions. Ongoing work explores the use of the C-Maps with other routing methods than traditional geographic routing, with same lightness considerations, to achieve even better routes.

Acknowledgments. This work has been supported by the Spanish CICYT DPI2006-15390 project and the GISED, group of excellence recognised by the Diputación General de Aragón.

References

1. L. Blazevic, J. Le Boudec, S. Giordano: A Location Based Routing Method for Mobile Ad Hoc Networks. IEEE Transactions on Mobile Computing 4(2) (2005) 97–110
2. P. Bose, P. Morin, I. Stojmenovic, J. Urrutia: Routing with guaranteed delivery in ad hoc wireless networks. Wireless Networks 7(6) (2001) 609–616
3. S. Chen, G. Fan, J. Cui: Avoid "Void" in Geographic Routing for Data Aggregation in Sensor Networks. International Journal of Ad Hoc and Ubiquitous Computing (IJAHUC), Special Issue on Wireless Sensor Networks, 1(4) (2006) 169–178
4. B. Karp, H. T. Kung: GPSR: Greedy Perimeter Stateless Routing for Wireless Networks. In Proceedings of the 6th ACM International on Mobile Computing and Networking (MobiCom 00) (2000) 243–254
5. Y. J. Kim, R. Govindan, B. Karp, S. Shenker: Geographic Routing Made Practical. In Proceedings of the USENIX Symposium on Networked Systems Design and Implementation (2005)
6. E. Kranakis, H. Singh, and J. Urrutia: Compass routing on geometric networks. In Proceedings of the 11th Canadian Conference on Computational Geometry (1999) 51–54
7. F. Kuhn, R. Wattenhofer, Y. Zhang, A. Zollinger: Geometric ad-hoc routing: Of theory and practice. In Proceedings of PODC (2003) 63–72
8. M. O. Legner: Map-based Geographic Forwarding in Vehicular Networks. University of Stuttgart, Faculty of Computer Science, Diploma Thesis No. 1994 (2002)
9. B. Leong, B. Liskov, R. Morris: Geographic Routing without Planarization. In Proceedings of the 3rd Symposium on Network Systems Design and Implementation (NSDI 2006) (2006)
10. M. Li, W.-C. Lee, A. Sivasubramaniam: Efficient peer-to-peer information sharing over mobile ad hoc networks. In Proceedings of the 2nd Workshop on Emerging Applications for Wireless and Mobile Access (MobEA 2004) (2004)
11. S. Ratnasamy, B. Karp, S. Shenker, D. Estrin, R. Govindan, L. Yin, F. Yu: Data-centric storage in sensornets with GHT, a geographic hash table. Mobile Networks and Applications 8(4) (2003) 427–442

A hardware accelerated implementation of the IEEE 802.15.3 MAC protocol

Daniel Dietterle, Jean-Pierre Ebert, and Rolf Kraemer

IHP microelectronics GmbH, Wireless Communication Systems, PO Box 1466,
15204 Frankfurt (Oder), Germany
{dietterle, ebert, kraemer}@ihp-microelectronics.com

Abstract. We present a hardware/software implementation of the IEEE 802.15.3 MAC protocol. Processing-intensive and time-critical protocol tasks are handled by a protocol accelerator that is integrated on-chip with a 32-bit general-purpose processor in order to achieve a moderate (20–40 MHz) system clock frequency. This enables low-power wireless devices compliant with this standard, providing high data rate, multimedia communication.

One of the main tasks of the protocol accelerator is to analyze received or transmitted beacons. Based on the channel time allocations broadcast in the beacon and frame information stored in a hardware transmission queue, frames are transmitted without immediate control of the processor. Other features of the protocol accelerator include CRC generation, handling of immediate acknowledgment frames, and direct memory access.

Keywords: Personal area networks, protocol implementation, hardware accelerator.

1 Introduction

Wireless communication systems have attracted tremendous research work in industry and academia in the past decade. As the outcome of this research, new communication protocol standards, applications and products have been developed. In the future, we will see new applications in, for instance, personal health care, multimedia entertainment, or industrial automation making use of wireless body area network (BAN), personal area network (PAN) or mesh networking technology. These applications are characterized by a demand for high data rates, quality-of-service (QoS) provisioning, as well as energy efficiency. Low energy consumption is critical for networks of battery-powered devices with targeted network lifetimes of months and years.

In 2003, the IEEE has standardized a medium access control (MAC) and physical layer specification for high data rate wireless PANs, known as IEEE 802.15.3 [1]. Work on improvements of the MAC layer, alternative physical layers as well as a mesh networking extension has continued. We see this standard as a good candidate for future wireless applications and an enabling technology for low-power, wireless multimedia communications for a number of reasons:

Please use the following format when citing this chapter:

Dietterle, D., Ebert, J.-P., Kraemer, R., 2007, in IFIP International Federation for Information Processing, Volume 248, Wireless Sensor and Actor Networks, eds. L. Orozco-Barbosa, Olivares, T., Casado, R., Bermudez, A., (Boston: Springer), pp. 215-226.

The IEEE 802.15.3 MAC protocol offers an isochronous data service, supporting multimedia as well as industrial applications with requirements for guaranteed transmission opportunities. To save energy, devices may go into power-save mode. Synchronized power-save sets ensure that all devices in the set wake up at the same time. The used channel access scheme is time-division multiple access (TDMA). This way, the protocol can easily be used with ultra wide band (UWB) transceivers, that do not provide a channel sensing capability. Contention access based on channel sensing is optional. Furthermore, the specified high data rates from 11–55 Mbit/s reduce the transmission duration of frames. While high-rate transceivers typically consume more *power* than low-rate transceivers, still the amount of consumed *energy* per bit can be less than for low-rate transmissions. High-rate transmissions become more energy-efficient when messages can be aggregated to larger packets and the overhead for switching the radio transceiver from power-down to active mode is reduced.

High data rates on the physical layer and complex protocol functions, such as scheduling allocated time slots and managing power-save sets, require more processing power on the devices, especially for devices that are capable of acting as piconet coordinator (PNC). However, in many applications all networked devices must be battery-powered, like in a wireless network of sensors located on the human body. We advocate the use of protocol accelerators for processing-intensive and time-critical protocol tasks in order to achieve a moderate (20–40 MHz) system clock frequency. With this approach, the protocol functionality is partitioned into software executed on a general-purpose micro-controller and hardware integrated on-chip with the processor.

The remainder of this paper is organized as follows. Section 2 briefly introduces the IEEE 802.15.3 MAC protocol with a stress on timing-critical functionality. In Sect. 3, our protocol design flow—from an abstract SDL model until hardware/software partitioning—is presented. In the main part of the paper, Sect. 4, the design of the protocol accelerator for the IEEE 802.15.3 MAC protocol is explained. Finally, we present related work and conclusions.

2   IEEE 802.15.3 MAC Protocol

In a wireless network operating according to the IEEE 802.15.3 standard, there is one piconet coordinator (PNC) and a number of associated devices. The PNC broadcasts beacon frames at regular time intervals. These beacons contain, among others, information about the time of the next beacon and when other devices may access the channel in the same superframe, i.e. in the time until the next beacon.

A channel time allocation list, which is part of the beacon frame, announces the time slots that are reserved exclusively for the devices. Additionally, the PNC may define a contention access period (CAP) following immediately after the beacon. In this CAP, all devices can access the channel using a random backoff procedure. Data communication among the associated devices is peer-to-peer. An example of a piconet and a beacon frame is shown in Fig. 1.

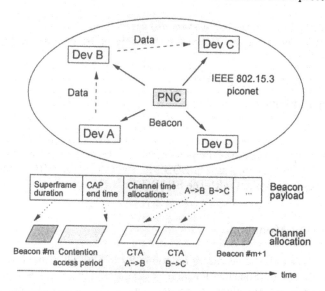

Fig. 1. IEEE 802.15.3 network topology and superframe structure (example)

IEEE 802.15.3 MAC frames consist of a 10-byte header followed by an arbitrary length payload field. Furthermore, a 32-bit frame check sequence (FCS) is calculated over the payload and added to all transmitted frames. The CRC-32 algorithm is used for this purpose. The frame receiver performs the same calculation and compares the final CRC value to determine if any bit error has occurred during the transmission.

In most cases, especially in the case of command frames, the sender expects the receiver to acknowledge the correctly received frame immediately, i.e. exactly 10 microseconds after the end of the frame. If an acknowledgement frame is not received, the sender will retransmit the frame. The basic MAC frame format and the immediate acknowledgment policy is shown schematically in Fig. 2.

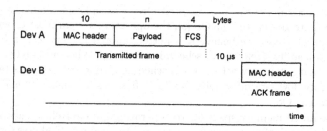

Fig. 2. IEEE 802.15.3 MAC frame format and immediate acknowledgment policy

The specified data rates in the standard range from 11 to 55 Mbit/s, however the same MAC protocol can be used with much higher or lower data rates, as well.

3   Design Flow

The starting point of our design flow was the IEEE 802.15.3 MAC protocol specification [1]. The protocol implementation is targeted for a system-on-chip based on the LEON2 processor. LEON2 is a synthesisable VHDL model of a 32-bit processor compliant with the SPARC V8 architecture [4].

3.1   Protocol Modeling

We have modeled all the necessary functionality in SDL [2] using Telelogic TAU SDL suite [3]. SDL (Specification and Description Language) is a formal description technique that allows to model behavior by means of communicating extended finite state machines. By extensively simulating our IEEE 802.15.3 MAC protocol model we could verify its correct functionality. Details of this model can be found in [10].

The validated SDL model is the basis for the MAC protocol implementation by an automatic transformation. The effort of re-implementing the protocol in C/C++ would be too high and error-prone compared to an optimization approach where inefficient SDL concepts in the model are replaced by equivalent functions with less overhead. Additionally, the time to achieve a fully tested implementation is considerably shortened.

3.2   Operating System Integration

The next step was to target the SDL model to an operating system (OS), in our case the Reflex OS [8]. For this purpose, we developed a so-called Tight Integration model [11]. This replaces the SDL run-time environment with a tailored, very efficient OS integration layer.

Reflex is a tiny, event-flow oriented OS for deeply embedded systems [7]. Although quite similar to TinyOS [9], the operating system most often used for wireless sensor nodes, we believe it is better tailored for our system because of its earliest-deadline-first process scheduling strategy. Whereas TinyOS tasks run to completion before any other task is scheduled, time-critical tasks (activities) will interrupt lower-priority activities in Reflex. Such a behavior is difficult to achieve in TinyOS.

The required memory space for the operating system Reflex, the integration layer, and a simple SDL system was measured to be about 20 kbytes for a system targeted for the LEON2 processor.

3.3   Hardware/Software Co-Design

Some of the MAC protocol functionality underlies tight timing constraints, for instance acknowledgment (ACK) frame transmission has to start exactly 10 microseconds after the end of a received frame. To identify bottlenecks in the pure software implementation and to estimate the required clock frequency to meet all timing constraints, we performed a profiling of the software. For that purpose, the software was simulated using the LEON2 instruction set simulator (ISS) TSIM [5].

Furthermore, with TSIM it is also possible to model the behavior of hardware components that are connected to the LEON2 processor via the on-chip bus. This enables the simulation of the system with protocol functionality mapped to hardware. To achieve this, the corresponding functions were removed from the software model and put into a hardware component. This allowed us to study the new timing behavior of the protocol implementation and optimize the hardware/software partitioning until all timing constraints were met and the required clock frequency was acceptable.

As a result of the hardware/software partitioning we identified the frame reception and transmission procedure, superframe timing control, immediate acknowledgment handling, and parts of the transmission queue to be designed in hardware. In other words, all the low-level, timing-critical and processing-intensive tasks of the channel access mechanism have been mapped to the hardware partition. This corresponds well to the lowest service layer in our SDL model, which is called Transport Engine (see Fig. 3).

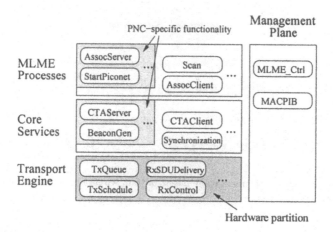

Fig. 3. Functional layering of the SDL processes of the MAC protocol model, partition mapped to hardware highlighted.

The remaining protocol functionality is handled by the LEON2 processor. Interrupts are used to signal protocol-related events from the hardware to the software. Conversely, the software interacts with the hardware block by writing to and reading from a number of control registers.

4   Protocol Accelerator Design

In this section we first introduce the target hardware platform for the protocol accelerator. Then, we present the overall architecture of the accelerator and how its components work together to provide the desired functionality. This is followed by a detailed presentation of the transmission queue component. We show how our design can be used for different data rates and non-standard physical layer timing parameters and how the protocol software and hardware accelerator interact with each other. Finally, the section is concluded with results from an FPGA implementation of the complete system.

4.1   Target Hardware Platform

The MAC protocol accelerator is one important component of our wireless communication platform. The heart of this hardware platform is the LEON2 processor [4], that runs protocol and application software. We pursue a single-chip integration of the MAC protocol, digital baseband processing and the RF frontend, as shown in Fig. 4.

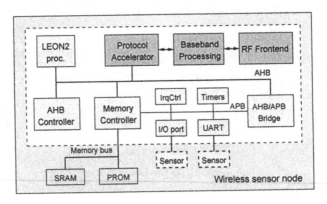

Fig. 4. Sensor node implementation based on our wireless communication platform (white box).

The protocol accelerator is connected to the system bus, the AMBA AHB bus (see Fig. 6). Via an AHB master interface it is possible for the accelerator to directly access the system memory, for instance to store and retrieve frame data without involving the LEON2 processor.

For data transfer to/from the baseband processor as well as status indications from the physical layer, the MAC-PHY interface was designed. It is of master/slave type with the MAC protocol accelerator acting as master. The accelerator sends commands to control the start of transmission or reception and to exchange frame data. The baseband processor contains a 32-byte data buffer for storing frame data in both, RX and TX, direction in order to decouple the two protocol layers with respect to timing (see Fig. 5). There are some signals from the physical layer to the MAC hardware accelerator that indicate the status of these buffers and can be used for flow control.

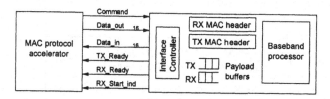

Fig. 5. Interface between MAC protocol accelerator and the physical layer. *TX_Ready* when TX buffer not full, *RX_Start_ind* when reception of new frame has started, *RX_Ready* when RX buffer not empty.

4.2   Architecture

Figure 6 shows the main components of the protocol accelerator. The tasks performed by each of the main components are listed below. They reflect the protocol functions that have been identified to be designed in hardware in Sect. 3.

- In receive direction, to retrieve frame data from the physical layer byte by byte, perform filtering and CRC check, and to store the data at a given memory location by means of direct memory access (components *Rx controller*, *CRC*, and *DMA*).
- In transmit direction, to retrieve frame data from a memory location, calculate and append the check sum, and to push the data to the physical layer (components *Tx controller*, *CRC*, and *DMA*).
- To signal a successful reception or transmission of a frame to the processor by an interrupt (component *Interrupts*).
- To analyze received and transmitted beacon frames and extract information on channel time allocations (component *Beacon parser*).
- To manage a queue of frames that are to be transmitted and to select an appropriate frame for transmission (component *Transmission queue*).
- At the start of a time slot or following a frame transmission, to query a new frame from the queue and, in the case that the frame must be acknowledged by the receiver, wait for the acknowledgment frame (components *Scheduler* and *Timers*).
- To perform the backoff procedure in the contention access period (components *Scheduler* and *Timers*).

- To send an acknowledgment at the right time upon reception of a frame that needs to be acknowledged (components *Scheduler*, *Timers*, and *Tx controller*).

An additional component (*CalcDuration*), that is not shown in Fig. 6 for simplicity, calculates the actual duration of a frame transmission based on its payload length and data rate. This component is used to determine if a frame transmission fits into an available time slot and when a transmission initiated by the protocol accelerator will be completed by the physical layer.

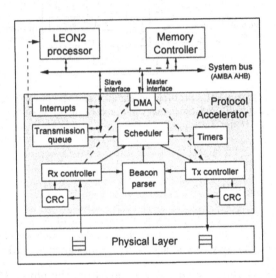

Fig. 6. Hardware architecture of the protocol accelerator (direct memory access data path highlighted).

4.3   Transmission Queue

The *Scheduler*, *Transmission queue*, and *DMA* components in the protocol accelerator facilitate a frame transmission operation that is not directly controlled by the processor. This allows to reduce the clock frequency of the processor and, hence, leads to possible energy savings. The *Transmission queue* component, that shall be discussed in this section, has a key role in this operation.

The transmission queue designed for the IEEE 802.15.3 MAC protocol accelerator contains and manages a table with information on frames to be transmitted—not the frame data itself, which is stored in application memory. The protocol software running on the LEON2 processor fills the table according to the generated frames and their transmission order. An interrupt is signaled to the processor as soon as a frame from the queue has been transmitted successfully or its maximum retransmission limit has been reached. This indicates to the software that the entry in the table is free and can be reused by another frame. An update, however, does not have to happen immediately as there are still enough frames in the hardware transmission queue.

The current design contains 8 table entries, but there might be many more, e.g. 32 or 64. In order to find the right frame upon request from the *Scheduler* quickly, there are ordered lists for different frame types, for instance for beacon frames or frames that can be transmitted in the contention access period. The table index of the first list item is kept in a register. From this item the complete list can be traversed by following the table index of the next list item, that is stored in the table. Furthermore, there is a reference to the previous list item to support delete operations efficiently. In essence, this forms a double-linked list. These lists are also used to preserve the right order of data frames where the MAC protocol provides this service to the higher layers.

4.4 Support for Flexible Timing

The protocol accelerator has been designed such that it is not limited to a fixed data rate or time intervals between consecutive frame transmissions (interframe spaces). Instead, a number of software programmable registers have been introduced that completely determine the timing behavior of the protocol. These registers are read by the *Scheduler* to calculate the point in time when the next frame can be transmitted. There are registers for SIFS, MIFS, and backoff slot duration, each given in microseconds.

Additionally, there is one programmable register which contains a rate factor and is used by the above mentioned *CalcDuration* component to calculate the duration of frame transmissions.

4.5 Software Interface

The protocol software interacts with the hardware accelerator through a set of registers that are accessible from the processor via memory mapped I/O. A special memory region is reserved for the accelerator. Additionally, interrupts are used to signal events from the protocol accelerator to the processor.

The protocol accelerator contains a number of configuration registers that store MAC protocol information like the piconet or device identifier. It is possible to enable or disable the scheduler with another register. When the scheduler is enabled, the protocol accelerator will analyze beacon frames and seize transmission opportunities, otherwise it is just in scanning mode and delivers received frames.

The protocol accelerator features four maskable interrupts: superframe start, MAC header received, MAC frame received, and transmission queue interrupt. The latter is triggered when a queued frame has been sent or was discarded. Two registers indicate the first free queue index and the first index that contains a transmitted or discarded frame. This way, the software does not have to browse the complete queue.

While the software is modifying the transmission queue, the hardware may not access its contents in order to avoid inconsistencies. Similarly, when the hardware is browsing the queue for a suitable frame to transmit, the software may not alter its lists. Therefore, a lock must be acquired by the software before an operation on the queue can be performed and released when it is done. When the queue is locked, the hardware cannot access the queue.

To control the receive operation, there are registers for the address where the next payload can be stored and for the size of this buffer. When the accelerator has written data to this buffer, no other payload can be received unless a new payload buffer address is provided by the software. An acknowledgment frame will be generated only if the header and payload of the received frame have been stored.

4.6   Results

In a first step, we have implemented the LEON2 system and protocol accelerator on a Xilinx Virtex-II FPGA. We have successfully tested the complete MAC protocol implementation, i.e. the protocol software running on the LEON2 processor and the protocol accelerator, by connecting two such boards with wires. This emulates a network of two devices. The wires couple the boards below the MAC layer, data symbols are transferred serially at a rate of 20 Mbit/s. Table 1 shows the usage of FPGA resources of the same LEON2-based system with and without the protocol accelerator.

Table 1. FPGA resources used by the MAC protocol system.

Resources	LEON2 system		Difference
	Original	With prot. acc.	
4 input LUTs	11,582	24,034	12,452
Occupied slices	6,828	14,365	7,537
Block RAMs	20	22	2
Equivalent gate count	1,427,060	1,681,651	254,591

After the successful test on the FPGA, the complete LEON2-based MAC processor including Flash memory and peripherals has been designed and taped out as an ASIC in 0.25 μm CMOS technology. The chip occupies an area of 31.9 mm^2 and consumes 15 mW/MHz. Based on our synthesis results, the silicon area of the protocol accelerator is about 1.8 mm^2.

5   Related Work

The design of dedicated hardware to speed up data processing and save energy has long been exploited in digital system design. MAC protocol accelerators for wireless communication systems have been reported, e.g. by Meng *et al* [13] and Fujisawa *et al* [14], for the IEEE 802.11a standard, and by Haroud *et al* [15] also for the IEEE 802.15.3 standard. In the latter publication, little is said about the mechanism to initiate frame transmission and the interface between hardware and software.

The use of hardware queues for autonomous transmission of packets is well-known practice and has also been described in [13] and [14]. However, hardware transmission queues are not limited to the design of wireless communication systems, but are used also in high-performance network switching devices such as ATM routers.

6   Conclusion

We have presented a hardware/software implementation of the IEEE 802.15.3 MAC protocol. A prototypical system has been developed for an FPGA board and shows correct and timely protocol behavior at a system clock frequency of 40 MHz with 20 Mbit/s physical layer data rate.

While details on our software design have been published elsewhere, we have focused in this paper on the design of a protocol accelerator and on the communication between the hardware and software part. Moderate clock frequencies of the general-purpose microprocessor—in our case the LEON2 processor—could be achieved mainly because of using direct memory access to payload data in RAM and a hardware transmission queue.

We plan to integrate hardware cores for encryption and decryption in the protocol accelerator, since security algorithms are data-processing intensive and also time-critical tasks. Our current hardware accelerator architecture allows this extension without much design effort.

The proposed protocol accelerator can be used as an IP core for future wireless multimedia communication systems that need to consume low power. The hardware provides programmable registers to adapt to the intended physical layer data rate and timing parameters. Furthermore, its standard AMBA AHB bus interface and the interrupt mechanism make it easy to be integrated with different hardware platforms. The software interface to the protocol accelerator is simple but effective.

Though the presented protocol accelerator is designed specifically to support the IEEE 802.15.3 MAC protocol, its architecture and many of its components can be reused with slight modifications for other protocols. In fact, we plan to use it for a low-power wireless sensor node based on the IEEE 802.15.4 standard, where the channel access mechanism is quite similar.

Acknowledgment

This work was partly funded by the Federal Ministry of Economics and Technology (BMWi) of Germany under grant no. 01 MT 306.

References

1. IEEE Standard 802, "Part 15.3: Wireless Medium Access Control (MAC) and Physical Layer (PHY) Specifications for High Rate Wireless Personal Area Networks," 2003.
2. ITU-T, "ITU-T Recommendation Z.100. SDL: Specification and Description Language," 1999.
3. Telelogic AB. (2004). Telelogic Tau SDL Suite [Online] Available: http://www.telelogic.com/products/tau/sdl
4. Gaisler Research AB. (2006). LEON2 Processor [Online]. Available: http://www.gaisler.com
5. Gaisler Research AB, "TSIM Simulator User's Manual," 2006.
6. Pender Electronic Design GmbH, "GR-CPCI-XC2V Development Board User Manual," 2005.
7. K. Walther, R. Hemmerling, and J. Nolte, "Generic Trigger Variables and Event Flow Wrappers in Reflex," in *ECOOP — Workshop on Programming Languages and Operating Systems*, 2004
8. J. Nolte, "Reflex - Realtime Event FLow EXecutive," Available from http://www-bs.informatik.tu-cottbus.de/38.html?&L=2, 2006.
9. J. Hill, R. Szewczyk, A. Woo, S. Hollar, D. Culler, and K. Pister, "System Architecture Directions for Networked Sensors," in *Architectural Support for Programming Languages and Operating Systems*, 2000, pp. 93–104.
10. D. Dietterle, I. Bababanskaja, K. Dombrowski, and R. Kraemer, "High-Level Behavioral SDL Model for the IEEE 802.15.3 MAC Protocol," in *Proc. of the 2nd International Conference on Wired/Wireless Internet Communications (WWIC)*, P. Langendörfer, M. Liu, I. Matta, and V. Tsaoussidis Ed. Lecture Notes in Computer Science, Vol. 2957. Springer-Verlag, Berlin Heidelberg New York, 2004, pp. 165–176.
11. G. Wagenknecht, D. Dietterle, J.-P. Ebert, and R. Kraemer, "Transforming Protocol Specifications for Wireless Sensor Networks into Efficient Embedded System Implementations," in *Proc. Third European Workshop on Wireless Sensor Networks (EWSN 2006)*, Lecture Notes in Computer Science, Vol. 3868. Springer-Verlag, Berlin Heidelberg New York, 2006, pp. 228–243.
12. D. Dietterle, J.-P. Ebert, G. Wagenknecht, and R. Kraemer, "A Wireless Communication Platform for Long-Term Health Monitoring," in *Proc. PerCom Workshops 2006*, 2006, pp. 474–478.
13. T. H. Meng, B. McFarland, D. Su, and J. Thomson, "Design and implementation of an all-CMOS 802.11a wireless LAN chipset," IEEE Commun. Mag., vol. 41, no. 8, Aug. 2003, pp. 160–168.
14. T. Fujisawa, J. Hasegawa, K. Tsuchie, T. Shiozawa, T. Fujita, T. Saito, and Y. Unekawa, "A single-chip 802.11a MAC/PHY with a 32-b RISC processor," IEEE Journal of Solid-State Circuits, vol. 38, no. 11, Nov. 2003, pp. 2001–2009.
15. M. Haroud, L. Blazević, and A. Biere, "HW accelerated ultra wide band MAC protocol using SDL and SystemC," in *Proc. IEEE Radio and Wireless Conference (RAWCON'04)*, IEEE, 2004.

A Proposal for Zigbee Clusters Interconnection based on Zigbee Extension Devices

Ángel Cuevas, Rubén Cuevas, Manuel Urueña, and David Larrabeiti

Departamento Ingeniería Telemática. Universidad Carlos III de Madrid
{acrumin,rcuevas,muruenya,dlarra}@it.uc3m.es

Abstract. WSNs are becoming an important topic in the research arena. Their low cost and the wide range of applications where wireless sensor networks can be applied, make them an important research and commercial field in the short term. Furthermore, the IEEE 802.15.4 standard for physical and MAC layer in addition with the Zigbee specification for the network and application layer, are an enforcement for WSNs. Much research in WSNs related topics such as energy-aware, MAC protocols, routing protocols, security, location, etc is being carried out nowadays. An open issue is the interconnection between wireless sensor networks where different approaches can be adopted: bridging (extension devices) and proxying (mainly middleware entities). This paper is focused in the bridging approach for Zigbee networks. Our objective is to establish a Zigbee Extension Device Network for the case where the different clusters can not be connected to the same Zigbee network by using the multihop routing protocol defined in Zigbee. A Zigbee cluster is a set of Zigbee nodes. This paper introduces three different solutions to join the disconnected clusters of the Zigbee Network: a central approach (based on the use of a central station) and two distributed approaches; one of them proposes that each Zigbee Extension Device stores the information of all the other Extension Devices within the network while the other uses a P2P Zigbee Extension Device Overlay Network to join the disconnected Zigbee Clusters. The paper includes the analysis of each solution and the comparison among them.

Keywords— **Zigbee, WSN, interconnection, cluster, overlay, p2p.**

1 Introduction

Recent advances in microelectromechanical systems, low power and highly integrated digital electronics, tiny microprocessors, low power radio technologies, etc, have permitted to develop many kinds of versatile sensors and actuators. These sensors can sense many environmental characteristics like temperature, humidity, pressure, speed, presence, etc. Due to the previous advances, low-cost and low-power wireless sensor are being installed nowadays. Some predictions claim that the price of a sensor will be lower than 1$. Therefore, it will be

Please use the following format when citing this chapter:

Cuevas, A., Cuevas, R., Urueña, M., Larrabeiti, D., 2007, in IFIP International Federation for Information Processing, Volume 248, Wireless Sensor and Actor Networks, eds. L. Orozco-Barbosa, Olivares, T., Casado, R., Bermudez, A., (Boston: Springer), pp. 227-238.

possible to develop networks with many unattended sensors. Typical applications for Wireless Sensor Networks (WSNs) are: military, environmental, health, home, environmental control in office building, managing inventory control,etc. Much research has been done in WSNs at all levels: physical, MAC, network and application layers. In [1] general concepts and classifications of the main characteristics within a WSN are studied. In [2] approaches for energy aware, routing, location and security are explained. Finally, [3] makes a taxonomy of the research approaches for routing in WSNs.

Efforts in standardisation have been carried out at different levels. IEEE 802.15.4 [4] is a standard for the physical and MAC layers. Furthermore, the Zigbee Alliance[1], formed by more than 150 partners, has developed an specification for the network and application layers, [5]. The Zigbee Specification operates over IEEE 802.15.4 MAC layer. The IEEE 802.15.4/Zigbee stack seems to be a suitable standard for the application development in WSNs environments.

This paper proposes a solution of an open issue of the IEEE 802.15.4/Zigbee stack. It is the necessity to interconnect WSN. In [6] two different approaches for WSNs (using the IEEE 802.15.4/Zigbee stack) have been identified:

Firstly, the interconnection of different WSN is described. For this purpose, it is necessary the presence of a *gateway* or *proxy*. The role of this proxy is to understand the messages coming from the WSNs, translate them into other messages and route them through a different network (e.g. IP backbone) to get to the target WSN. This proxies are proposed as complex elements which understand application level, so that each application needs a different proxy. [7] proposes a middleware approach using a P2P overlay and UPnP proxies to discover services.

Secondly, the interconnection of clusters (set of Zigbee nodes) is introduced. It is possible a scenario where different clusters of the same Zigbee network were not reachable among them by using multihop wireless routing mechanisms due to the physical distance. In that case, the Zigbee clusters must be interconnected by using another network infrastructure. In this case, the entity which attaches the Zigbee cluster to the infrastructure and allows the communication with other Zigbee clusters (ZCs) is called Zigbee Extension Device (ZED).

Our paper is focused on the second approach, the interconnection of ZCs belonging to an unique Zigbee Network. For this purpose, several architectures are proposed.

Firstly a central architecture is defined. In this approach, all the information about each ZC (i.e., the Zigbee addresses within the cluster and the Cluster ZED) is stored in a Central Station. A ZED uses the central station when inter-cluster communication is needed in order to obtain the IP address of the ZED which is managing the destination ZC.

Moreover, the paper describes two distributed architectures. In the first one each ZED knows all the information about the rest of ZCs (i.e. the ZED and the Zigbee addresses). Therefore, when a ZED receives a Zigbee command for an external Cluster, it searches locally the destination ZED and sends to it

---

[1] http://www.zigbee.org

the command. The other distributed solution is based on peer-to-peer (p2p) technology. Unlike the previous one, the ZEDs locates the target ZED for a certain Zigbee command by launching a search on a p2p overlay formed by all the ZEDs.

The remainder of the paper is organised as follows: In section 2 we present a background of IEEE 802.15.4/Zigbee stack focused mainly in the network layer. In section 3 our Zigbee Extension Device Network proposal is introduced. We analyse and compare the proposed architectures, in section 4. Finally conclusion and future work is presented in section 5.

2   Background

2.1   IEEE 802.15.4

IEEE introduced the standard IEEE 802.15.4 [4], for low rate wireless personal area networks (LR-WPAN). It standardises both physical and MAC layer. The advantages of LR-WPANs are: easy installation, very low-cost, reasonable battery life, reliable data transfer and a flexible protocol stack.

2.2   Zigbee

The Zigbee Alliance introduces in [5] a specification which defines the Network and Application layers working over the IEEE 802.15.4 stack. A good summary of Network and Application layers is shown in [2].

The Zigbee stack is built on top of the IEEE 802.15.4 stack. Network Layer functions are: multihop routing, route discovery and maintenance, security and joining/leaving a network. The network level is also in charge of assigning 16 bit addresses to the newly joined devices.

Three types of devices are defined by Zigbee: end-device, router and coordinator. The first one works as a sensor/actuator. A Zigbee router is a device with routing capabilities. Finally, a Zigbee coordinator is the one which creates the network and manages the network and there is only one per Zigbee Network.

A network address has 16 bits. Therefore, networks up to 65.536 nodes could be developed. Three network topologies could be defined at this level: star, tree and mesh. In the first one, all nodes are directly connected to the coordinator. In the tree topology, a tree is formed with the coordinator as root, and with the end-devices as leaves. Routers are used in this topology to join the different levels in the tree structure. The last topology is a free mesh topology where end-devices can communicate among them by using multihop routing through the routers.

Next, several network layer functions are explained.

2.3   Network Formation and Address Assignment

When a node wishes to join the network, it runs a join procedure. The network layer of this node starts a network discovery procedure. By using the MAC level the network layer can discover the routers within its radio coverage which are announcing their networks. It is function of the higher layers to decide which network to join. Then, the network layer selects a parent node from those available in the selected network and ask the MAC layer to start an association procedure. The parent node assigns the new node a 16 bit address. Therefore, in a Zigbee network a relationship parent-child is present in the address assignment. In [5] a distributed algorithm is defined to assign addresses. In this algorithm, the Zigbee coordinator sets the maximum number of routers (R_m), end devices (D_m) that each router may have as children and the maximum depth of the tree (L_m). A new router joined to the network receives a range of consecutive addresses based on its depth in the tree. From this range the first one is assigned to the new router while the others can be assigned for this router to their children. These children could be end-devices or other new routers. Nodes at depth Lm are assigned a single address. Also a single address is assigned if the new node within the network is an end-device.

In [2],the mathematical formula to calculate the address range for a router and the method to assign zigbee addresses are explained in detail.

2.4 Routing

Since several topologies are possible, different algorithms are applied depending on which is the current topology.

In a star topology it does not make sense talk about a routing protocol because all the messages are sent to the coordinator.

In a tree topology a tree-based routing algorithm is defined. An end-device always sends the messages to its parent which is a router (or coordinator). When the router receives this message it checks if the destination address is included in its range of addresses. In this case the router is able to route the message to the suitable end-device or router. Otherwise, the router forwards the message to its parent

In the mesh network topology the routing algorithm is more complex. This algorithm is based in Routing Tables (RTs) which can be constructed and updated by using a route discovery algorithm. This mechanism is described in more detail in [2]

2.5 Route Discovery

This process is required in order to create and update routing table entries in the routers and coordinator.

The Route Discovery mechanism is based on Ad Hoc On Demand Distance Vector (AODV) routing algorithm which is explained in [9] Basically, when a node needs a route to a certain destination broadcasts a Route Request message (RREQ). This RREQ travels along the network. The RREQ message accumulates a cost through the path it follows. This cost can be incremented by one in each hop, or other metrics can be used. When the destination node receives the RREQ, it replies a Route Reply (RREP) which travels back along the same path.

The RREP also carries information about the path cost which is incremented hop by hop by the node which forwards the message. Therefore, the originator node will receive several RREPs and it will chose as next hop the node which give back to it a minimum path cost to the destination node. Each intermediate node in the path will do the same, and by doing so the path is established. This process is explained in detail in [2].

3   Zigbee Extension Devices Interconnection Network Architectures

The Zigbee alliance identifies the necessity to interconnect [6] Zigbee Clusters (ZCs), which are a section of a Zigbee network formed by one or more devices, that can not communicate among them using routing mechanisms defined in the Zigbee Specification. To do this Zigbee bridges, which are called Zigbee Extension Devices (ZEDs), are introduced [6]. ZEDs function is to encapsulate the Zigbee stack used in the ZEDs interconnection network which could be based in many technologies (e.g. IP, Ethernet, WiFi....). In this paper is proposed the use of an IP backbone using the stack shown in figure 1.

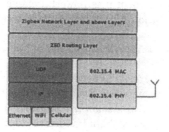

Fig. 1. ZED protocol stack

Thanks to this ZED, a single Zigbee network can be formed by joining ZCs. Thus, there is only one Coordinator within the network and a common address space is used. Typically, the ZEDs joined to the interconnection network are Zigbee Routers, but it is possible that Zigbee End Devices become ZEDs if they implement the interconnection protocol stack, however usually these end devices are very constraints nodes.

We propose three different architectures which are shown in figure 2, to solve the interconnection of ZCs. The principal objective is not to change anything in the current commercial Zigbee devices, but only introducing the ZED provide a full Zigbee connection through the interconnection network.

Two different scopes can be discussed in the proposed solution:

- **Intra-ZED routing:** It is the routing among Zigbee nodes located in the same ZC. It could be based-tree routing or mesh routing. The proposed architectures work with both of them.

– **Inter-ZED routing:** It is the mechanism to communicate ZEDs. The ZEDs have to solve the mapping between Zigbee destination addresses and IP ZEDs addresses, it means the ZED IP address which interconnects the ZC where the destination Zigbee node is located (if the backbone was not an IP backbone, a different mapping would be necessary). The mapping mechanism is different in each one of the proposed architectures. After discovering the destination ZED IP address all the solutions proposed do the same: the Zigbee message received is encapsulated into the UDP/IP stack and it is sent to the destination ZED, this one decapsulates the Zigbee message and forwards it into its ZC.

Following, the proposed architectures to solve the previous issues will be introduced:

1. **ZEDs connected to a central station:**
   This is the most intuitive solution. This central station has to store information about the whole Zigbee network. The minimum information stored in the central station for each ZC is the set of Zigbee addresses assigned to that ZC and the IP address of the ZED which connects that ZC to the Zigbee Network. Extra information could be stored. This central station could be placed in the coordinator if it is powerful enough. Another possibility is to use a monitoring station as a suitable place to allocate our central station (see figure 2).

2. **ZEDs storing information about all other ZEDs within the Zigbee Network:**
   In this approach every ZED has to store information for all the others ZEDs in the interconnection network. Each ZED stores the same information which was stored in the central station in the previous case, the IP addresses of all the ZEDs and the set of zigbee addresses located in the ZC associated to each ZED. In this architecture, the ZEDs form an overlay network. It is a distributed solution (see figure 2).

3. **P2P ZEDs Overlay Network:**
   The last approach introduces the use of P2P Structured Overlay Network in order to interconnect all the ZEDs within the Zigbee network. The proposed P2P network type is a fully distributed, self-organised, structured and searching based on Distributed Hash Table (DHT). All the standard mechanisms of these P2P networks, introduced in [8], could be used: joining procedure to the p2p, distributed storage, searching methods, etc (see figure 2)..

Some issues posed by this approach must be solved in the proposed architectures. The main issues are: join procedure (Zigbee address assignment), routing unicast inter-ZEDs (mapping between Zigbee addresses and IP addresses) and broadcasting inter-ZEDs.

In the join procedure there are two steps. Firstly, a new ZED joins the ZEDs interconnection network receiving the ZEDs IP address of its parent within the Zigbee address tree. This process is solved by different methods in each architecture (see subsection 3.1). Secondly, the new ZED joins the Zigbee Network

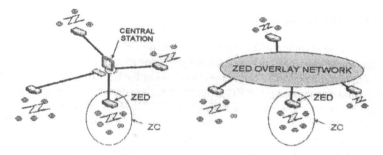

Fig. 2. Proposed architectures to interconnect ZCs by using ZEDs. On the left the scenario of ZEDs connecting to a central station is shown. On the right a figure for the distributed architecture is shown.

sending a Zigbee join message through an UDP/IP tunnel to its parent. This second step is common to all solutions.

In the routing unicast, when a ZED (ZED1) receives a message directed to a Zigbee node which is not located in its ZC, ZED1 has to find the IP address of the destination ZED (ZED2) which is connecting the ZC where the Zigbee destination node is located. This mapping process Zigbee address/IP address is solved using different mechanisms in each proposed architecture (see subsection 3.2). From this point, in all the solutions, the ZED1 sends the Zigbee message to the ZED2 using the IP backbone. The ZED2 decapsulates the message and forwards it into its ZC.

Following Zigbee address assignment, address mapping and broadcasting will be explained. In the next section, advantages and problems of each approach will be explained and also a comparison and suitable scenarios of applications for each one in front of the others is made.

3.1   Zigbee address assignment

In the central station architecture, a new Zigbee ZED has to obtain a position in the Zigbee address tree. In order to complete this process, the new ZED sends a query to the central station asking for a parent, and the central station returns the IP address of the ZED which will be its parent in the Zigbee address tree.

The joining procedure presents a problem in the second approach, because a new method has to be defined when a new ZED wants to join with the Zigbee Network. In somehow, the new ZED, ZED1, has to discover other ZED, ZED2 (e.g. a bootstrapping server could be used).

In the P2P approach, the new ZED follows a two step process. Firstly, the ZED joins the P2P network using a standard P2P mechanism and once it has joined to the P2P network the ZED joins the Zigbee Network. After the ZED has joined the P2P network, it has a number of neighbours, which are other ZEDs. Therefore, the ZED could use these ZEDs to join the Zigbee Network. The new ZED selects one of its p2p neighbours as parent in the zigbee address tree.

3.2 Address mapping

In the central station architecture, eventually, one ZED (ZED1) receives a Zigbee message whose destination is a node which is located in a different ZC. Then, ZED1 sends a query to the central station asking for the IP address of the ZED (ZED2) that is managing the ZC where the zigbee destination node is located. The central station answers to ZED1 giving the ZED2 IP address.

In the second approach the mapping is immediate, when a ZED (ZED1) receives message directed to a zigbee node located in a different ZC, the ZED1 checks in its information table which is the ZED which is managing the ZC where the destination node is located (ZED2).

Finally in the P2P approach, all ZEDs registered in the P2P have to publish its IP address and the set of zigbee addresses in the ZC attached to that ZED. Then, when data message arrives to a ZED (ZED1) and this one checks that the destination node it is not located into its ZC, the ZED1 launches a search query within the P2P using as key the Zigbee destination address. As result ZED1 obtain the IP of the ZED (ZED2) which has the destination node into its ZC

In the central station and P2P an improvement could be developed. When a mapping is established, the current ZED could keep this information into a cache. For example, this is very useful in an application when a sensor/actuator in a ZC communicates frequently with a sensor/actuator located in a different ZC. If the communication uses ACKs, the destination ZED could use a cache as well.

3.3 Broadcasting

In the central station solution, broadcasting is very simple. A ZED receiving a zigbee broadcast message encapsulates it into an UDP/IP packet and sends it to the central station. The central station checks that it is a broadcast message and forwards it to all the ZEDs except the originator ZED. Every ZED receiving the IP packet decapsulates the Zigbee broadcast message and forwards it into its ZC.

In the second architecture, the ZED receiving the zigbee broadcast message, identifies that it is a broadcast message and encapsulates it into the UDP/IP stack. After that, it sends IP packets to all the ZEDs stored in its information table. Each ZED receiving this packet decapsulates the zigbee broadcast message and broadcasts it into its ZC.

In the ZED P2P Overlay Network the broadcasting does not have an intuitive solution as in the previous architectures. When a ZED (ZED1) receives a zigbee broadcast message, it identifies the broadcast message and encapsulates into an UDP/IP packet. The ZED1 sends this packet to the nearest neighbour, this one does the same and it decapsulates the zigbee broadcast message and broadcasts it into its ZC. By doing so, the broadcast message reaches all the nodes into the ZEDs P2P overlay network. For instance, in a P2P ring structure with N nodes, it is necessary to forwards N times the broadcast message.

A better mechanism would be that the ZED1 would send the broadcast packet to all the nodes which it knows in the P2P. In reception of this packet, a ZED (ZEDi) checks if it has received the same packet before (it is necessary to identify the current broadcast message in the P2P overlay and it is not enough with checking that it is a Zigbee broadcast message). If it is true the ZEDi drops the packet. Otherwise, first the ZEDi forwards the IP packet to all the ZEDs that it knows, next, it decapsulates the zigbee broadcast message and broadcasts it into its ZC. All the ZEDs in the P2P repeats the same operation, so that the broadcast message reaches all the nodes into the ZEDs P2P overlay network.

RREQ is a special broadcast message which is sent to a target node. In a typical Zigbee network it is boradcasted everywhere because it is not known where is located the destination node. However, in our proposal, if a ZED receiving a RREQ is able to identify that the broadcast message is a RREQ, it could check the destination Zigbee address, and sends it to the suitable ZED, instead broadcasts the RREQ.

4   Analysis and Comparison of the proposed architectures

4.1   Analysis

The central station approach is a simple and intuitive approach. The principal problem of this solution is that it becomes a unique point of failure. If the central station fails all the network is disconnected and only local ZCs communications are available. Also, all the network information is located in a single point. Hence, if that information is lost the network is disconnected as well. Obviously, the solution to solve these problems is to replicate the information or to use a backup central station just in case the main central station fails. Other problem in this scenario is that some administration is needed. When a new ZED is introduced the information of this ZED must be updated in the database located in the central station. By using a simple protocol (register message) this task could be automatised.

The second approach has several disadvantages. First, a new mechanism must be defined in order to permit new ZEDs to join the Zigbee network. Basically, this mechanism should permit the new ZED to discover another ZED which would become its parent in the zigbee address tree. A bootstrapping server is an immediate solution for this problem another option could be manual configuration in order to join a new ZED the Zigbee Network. Another problem occurs when a new ZED joins the network because the information of this new ZED must be updated in all the nodes within the overlay network, and if the number of ZED is high it could be a hard administration task. On the other hand, the resolution of the IP address of the destination ZED is faster than in the other approaches because third entities are not needed (an improvement using caches for the other solutions has been introduced in 3.2). This is a distributed architecture but the information storage is not distributed.

Finally the ZED P2P Overlay Network has a main disadvantage in front of the other two, which is the delay in the communication because the time used to find the destination ZED IP address is longer than in the two previous cases. A basic analysis of this will be done in 4.2. Advantages for this architecture is that the ZED P2P Overlay Network is self-configured (the central station approach could be self-configured as well, if a simple register protocol is defined), the information is distributed and replicated and the ZEDs needs little storage capacity (an analysis of the storage capacity in the three architectures will be done in the 4.2), furthermore all these features are given by standard P2P mechanisms.

4.2   Comparison

The central station approach presents a centralised element which usually is a node with a high storage and processing capacity. Therefore, this additional processing capacity could be used ,for instance, to allocate the new ZED in a suitable position within the zigbee address tree.

In contrast to this central approach, the second and third architectures are distributed. Therefore, in the central station approach a unique point of failure is present. This problem could be solved by using a backup central station, but it supposes an extra cost element. This problem is not presnet in the distributed solutions.

In the second approach, ZEDs storing information about all other ZEDs within the Zigbee Network, has a problem in the joining procedure as we explained in 4.1. In the central station and P2P Overlay Network approaches how a new ZED joins the network has been explained in 4.1. Also the second approach is a worse solution than the other two when new ZEDs are added to the Zigbee Network. When a new ZED is added in the second approach, all the other ZEDs have to update their information tables with a new entry for the new ZED. If the number of ZEDs is large, it could be a problem. In the central station approach, only the information stored in the central node needs to be updated. In the P2P architecture, the new ZED has to publish the suitable information in the P2P and it is stored into another ZED by using P2P standard mechanisms.

Therefore, the P2P is self-configured by using P2P standard mechanisms, whereas the other two approaches are not. Obviously, these ones could be transforming in self-configured by defining some mechanisms, but these mechanisms are not standard as in the P2P.

If we do a comparison in terms of delay the best architecture is the second one. We define t as the time used for a communication from ZED to ZED or from ZED to central station and we compare the time to communicate from a ZC1 (ZED1) to another ZC2 (ZED 2). Table 1 shows in the second and third columns a comparison among the delay for each solution to discover the destination ZED (ZED2) for unicast and broadcast respectively. It must be noted that N is the number of ZEDs and the maximum number of hops in a P2P to find the peer storing the desired information is $log_2(N)$ [8].

It must be noted that the real delay will depend on the IP location of the different elements.

Some mathematical analysis can be done to see the information storage in the different approaches. The fourth column in table 1 establish a comparison

	Unicast Delay	Broadcast Delay	Storage Cost
Centralised Solution	$2*t$	$2*t$	N
Distributed Solution	0	t	N^2
P2P Solution	$log_2(N)*t$	$N*t$	k

Table 1. Metrics comparison

between the number of entries which must be stored in each solution. In the central station the value is referred to the central station, whereas in the distributed solutions the values represents the number of entries stored in each ZED. In the P2P solution, in order to get fault tolerance redundancy must be added. Therefore, each entry is stored in k peers. It must considered that each solution has extra storage due to the interconnection network used. For instance, in the P2P solution each ZED has to store information about its neighbours.

4.3   Suitable scenarios for each architecture

If the scenario presents a monitoring central station, something typical in wireless sensor networks, the most suitable architecture is the central station. Operation and management are also simpler form a central station. Furthermore, if the coordinator is a node with high processing and storage capacity it could be used as central station.

The architecture where ZEDs store information about all other ZEDs within the Zigbee Network is applicable in static scenarios where usually new ZEDs do not join the Zigbee Network. It is also applicable to small scenarios, where the information stored in each ZEDs is not too much. This architecture is worse than the others in scenarios where many ZEDs are needed and where new ZEDs are joined frequently.

The P2P ZEDs Overlay Network architecture is suitable in scenarios where high availability is needed (e.g.warfare). Also, in very large WSNs where delay can be trade-off with memory and robustness in unstable clusters.

In special scenarios where ZEDs are joining and leaving frequently (e.g. mobile ZC attaching the ZED network at different locations) the most suitable solution is to use the central station approach if a register protocol has been defined, otherwise the P2P solution would be the best solution.

5   Conclusions and future work

WSN interconnection is an open research issue. The Zigbee alliance [6] defines two possible approaches to solve this problem. The first one is the usage of proxies or gateways. Some research is being carried on in this field, and middleware based approaches have been presented in the literature. This paper is focused on a second approach which is the use of bridges or expansion devices . In particular we have proposed and analysed some Zigbee Extension Device

Interconnection architectures in this paper. Our aim has been to create a single Zigbee network with Zigbee Clusters that are not reachable among them by using standard Zigbee multihop routing protocol and they are members of the same Zigbee Network. We have presented the main issues to solve in order to be able to operate without changing nothing in the commercial Zigbee devices which implement the IEEE 802.15.4/Zigbee stack. With this objective in mind, three approaches has been introduced. Advantages and problems for each architecture have been discussed.

Further work is to implement a ZED into a Zigbee mote and start to develop the centralised model due to its simplicity. In particular, we want to study the case when the interconnection network is a cellular network (e.g 3G) where the setup and communication cost must be minimised.

Acknowledgement

The authors would like to thank the anonymous reviewers for insightful ideas and suggestions that made it possible to write a better article.This paper has been partially granted by the Spanish Government through the IMPROVISA project (TSI2005-07384-C03-027) and the regional government of Madrid through the BIOGRIDNET project (S-0505/TIC-0101).

References

1. I.F. Akyildiz, W. Su*, Y. Sankarasubramaniam, E. Cayirci. "Wireless sensor networks: a survey", Computer Networks 38 (4) (2002) 393-422.
2. Paolo Baronti, Prashant Pillai, Vince Chook, Stefano Chessa, Alberto Gotta, and Y. Fun Hu. "Wireless Sensor Networks: a Survey on the State of the Art and the 802.15.4 and ZigBee Standards", Technical Report, May 2006.
3. K. Akkaya, M. Younis, "A survey on routing protocols for wireless sensor networks", Ad Hoc Networks 3 (3) (2005) 325-349.
4. Institute of Electrical and Electronics Engineers, Inc., IEEE Std. 802.15.4-2003. "Wireless Medium Access Control (MAC) and Physical Layer (PHY) Specifications for Low Rate Wireless Personal Area Networks (LR-WPANs)", New York, IEEE Press. October 1, 2003.
5. ZigBee Alliance. "ZigBee Specifications", version 1.0, April 2005.
6. Patrick Kinney. "http://www.zigbee.org/en/documents/SensorsExpo/7-Sensors-Expo-kinney.pdf"
7. Manabu Isomura, Till Riedel, Christian Decker, Michael Beigl, Hiroki Horiuchi. "Sharing sensor networks", Proceedings of the 26th IEEE International Conference on Distributed Computing Systems Workshops (ICDCSW'06), July 2006, pp. 61-66.
8. J. Risson and T. Moors. "Survey of Research towards Robust Peer-to-Peer Networks: Search Methods", Internet Draft, draft-irtf-p2prg-survey-search-00.txt J. Risson and T. Moors: "Survey of Research towards Robust Peer-to-Peer Networks: Search Methods", Accepted to appear in Computer Networks
9. Charles E. Perkins and Elizabeth M. Royer. "Ad-hoc On-Demand Distance Vector Routing", Proc. 2nd IEEE Workshop on Mobile Computing Systems and Applications (WMCSA 1999), pp. 90-100, New Orleans, LA, USA, February 1999.